ENTROPY AND SUSTAINABILITY

ENTROPY AND SUSTAINABILITY

Michael Köhler

JENNY STANFORD
PUBLISHING

Published by

Jenny Stanford Publishing Pte. Ltd.
101 Thomson Road
#06-01, United Square
Singapore 307591

Email: editorial@jennystanford.com
Web: www.jennystanford.com

British Library Cataloguing-in-Publication Data
A catalogue record for this book is available from the British Library.

Entropy and Sustainability: Challenges of Future Energy and Material Management from a Thermodynamic Point of View

ISBN 978-981-5129-99-1 (Hardcover)
ISBN 978-1-003-74889-2 (eBook)

DOI: 10.1201/9781003748892

Contents

Preface

The responsible use of our planet's resources is the central question of humanity's survival. What needs to be done to overcome the overexploitation of natural resources and to develop an energy and materials economy that preserves living conditions on Earth and ensures sustainable development? There is a central thermodynamic quantity that has to be reconsidered for all future activities—this is entropy. This book introduces the physical basics and the historical scientific background of the entropy concept. It makes clear the role of entropy for living beings, machines, computers, industrial facilities, ecological systems, and Earth as a whole. It explains the necessity of the production of entropy, on the one hand, and the required possibilities for a sustainable entropy export, on the other.

The book is for everyone who is interested in the environment, sustainability, and its feasibility from a scientific point of view. It addresses the necessity to understand the fundamental importance of entropy in natural and technical processes as a basis for achieving an intelligent entropy management, which allows converting the different recent competing production processes into a global sustainable economy.

Acknowledgement

The book is an improved and extended version of the German version, *Entropie-Wende*, written and published a few years ago. For discussion and several important suggestions, I cordially thank Reinhard Guthke (Jena) and Andrea Knauer (Ilmenau). I would like to dedicate this book with gratitude to my longtime colleagues Alex, Andrea, Frances, Mike, Steffen, and Tina for all kindness and cooperation.

Introduction

Always, people have worked for a secure future. It was about securing the livelihood of the family, the economic future of the children, the profitability of a business, the long-term security of supply, and the safety of cities and countries. The desire to preserve what has been achieved and to create opportunities for further favorable development was paramount. To achieve this goal, natural resources had to be developed and utilized. The better this was achieved, the more promising it was to secure the future. Therefore, new ways were constantly sought to use and exploit natural resources, creating favorable conditions for future security and competition.

However, natural resources are limited, and their exploitation and overuse repeatedly reach limits, thereby working against the intended security of livelihoods.[1] This problem is tightened by the growing human population.[2] The deterioration of conditions under which resources can be used with increasing intensity or duration of use is a typical example of negative feedback. The goal of long-term future security must be to avoid such negative feedback.[3]

[1]D.H. Meadows et al., 1972.
[2]Scenarios of population development: Brook, B.W. et al., 2021.
[3]D. Meadows et al., 2006.

Entropy and Sustainability: Challenges of Future Energy and Material Management from a Thermodynamic Point of View
Michael Köhler

ISBN 978-981-5129-99-1 (Hardcover), 978-1-003-74889-2 (eBook)
www.jennystanford.com

DOI: 10.1201/9781003748892-1

In recent decades, it has become evident that limitations in the use of natural resources not only have local effects but also have massive global consequences. Not only are the sources of crude oil limited, but also their exploitation demands increasingly higher efforts. In addition, the massive global use of fossil fuels is problematic and dangerous for the climate and the stability of life-friendly conditions on the Earth and can cause a very dramatic development in the next centuries.[4] Therefore, there should be a transition in energy production leading to a fully sustainable provision of energy on Earth. This process is connected with the term "Energie-Wende" (energy transition).

Correcting mistakes with global impacts is not an illusion. The handling of the ozone hole is an encouraging example for humanity, showing that it is possible to correct human-caused global ecological missteps. For instance, the worldwide ban on the mass use of chlorofluorocarbons has been able to slow down their destructive effect on our atmospheric UV shield, and it is hoped that the stratosphere will gradually further regenerate over the next decades.

Worldwide, economic, political, and scientific-technical efforts are being made to replace the largely unsustainable energy production with environmentally friendly, sustainable processes. On one hand, the extraction of fossil fuels is becoming increasingly complex and risky, while on the other hand, the use of coal, oil, and natural gas leads to the continuous accumulation of carbon dioxide in the atmosphere, resulting in growing dangers associated with an enhanced global greenhouse effect. The consumption of fossil energy resources and the release of greenhouse gases are causally linked. They are two sides of the same coin. The conversion of coal, oil, and natural gas into combustion products is a prerequisite for the production of electrical energy from chemical energy and its use in the form of fuels. The first law of thermodynamics states that energy cannot be created from nothing but can only be transformed. In the case of fossil fuels, the chemical energy stored in the deposits is made usable in a heat-releasing, i.e., "exothermic," process and converted into

[4] Lyon, Ch. et al., 2021.

thermal, kinetic, and electrical energy. Without energy sources, no energy can be provided. Fossil fuels have been the dominant energy source for about two centuries. We owe the development of industrial society and today's high average standard of living in industrial societies to them. The price for their use is the exploitation of natural deposits and the excessive release of carbon dioxide into the atmosphere.

A way out seems to be provided by methods that use so-called renewable energy sources. Hydropower, wind, and solar energy are among them, as well as energy production from sewage gases and agriculturally produced biomass. Directly or indirectly, solar radiation ensures the provision of this energy. Even in the generation of energy from wind, hydropower, and biomass, energy production is linked to the current energy flow from the sun. The sun drives atmospheric convection and is thus the cause of the wind. It causes seawater to evaporate, allowing clouds to form, which, especially in mountains, rain down and feed rivers and reservoirs, and it supplies green plants with light for photosynthesis. In all these processes, the energy comes indirectly but also temporarily from the sun. In these processes, energy conversion is closely linked to material processes. Only in photovoltaic solar systems is electrical voltage generated directly from the absorbed light energy of solar radiation. An alternative energy source is heat from the Earth's interior, whose origin is mainly due to natural nuclear processes and tidal forces. The economically sensible use of "geothermal energy" requires larger temperature differences and comparatively rapid heat transfer processes between the local heat reservoir in the Earth's crust and the atmosphere, a condition that is met, for example, in volcanically active areas.

Behind the new paths in energy supply lies the concept of sustainability. It means that resources should only be consumed to the extent that they can be regenerated by natural processes. This principle is massively violated in the use of fossil fuels such as coal, oil, or natural gas. In contrast, the light for photovoltaic systems and plant growth for biogas production, and the energy for wind and hydropower are constantly replenished by solar radiation. Therefore, these energy production strategies are also referred to as regenerative.

The concept of sustainability originates from forestry. Central European forests were increasingly overexploited from the Middle Ages to the early modern period due to logging, forest grazing, litter extraction, and particularly the enormous demand for charcoal in metallurgy. To curb this development, around 1700, the principle of sustainable forestry was introduced. It states that no more wood may be harvested than can regrow. The forestry idea was to have forest sections at various stages of growth so that wood could be harvested at any time, while ensuring that wood could also be harvested in the future.[5] Overall, this was intended to achieve a quasi-stationary state while simultaneously utilizing wood (Fig. I.1).

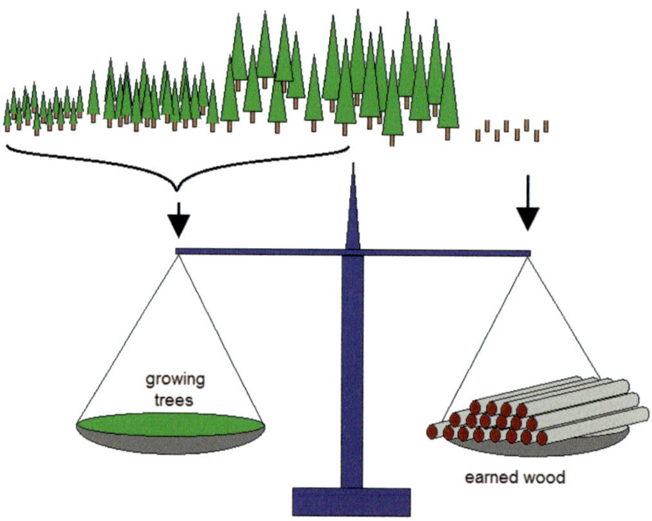

Figure I.1 The principle of sustainable forestry.

The problem of forest over exploitation and its solution through sustainable management can be fundamentally applied to the use of other limited resources. To maintain a current state for the future, generally, no more resources should be extracted than are produced by natural processes. The demand for

[5]H.C. von Carlowitz, 1713.

sustainability concerns all materials on the one hand, but also the production of energy on the other.

The effort to set up wind turbines, solar panels, and hydropower plants is considerable, but their operation is less costly and labor-intensive than conventional power plants. However, these facilities are often associated with environmental impacts—citizen protests against new dams, power lines, and wind farms make this clear—and they have only recently become economically competitive with the construction of fossil fuel power plants, whose indirect costs to the environment and health are usually excluded from profitability calculations.

Thus, the energy transition not only means expanding new and alternative energy generation facilities but also rethinking energy and energy transformation: Are we using energy correctly? Is energy fundamentally conserved? How are material and energy transformations connected? To what extent is the Earth gradually heating up? Can we utilize waste heat in energy production? What happens to the heat into which all energy is eventually transformed? Why can energy be transformed at all? How does energy transformation work in energy production and consumption? How can sustainability be ensured in energy production, material management, and food production?

These questions encompass numerous scientific and technical individual problems. Mechanisms need to be understood in detail to achieve improvements. Technical, economic, chemical, physical, and biological factors all play a role. However, there is one central thermodynamic quantity that connects the processes in all these fields. That is entropy.

Chapter 1

Concept of Entropy in Thermodynamics and Physical Chemistry

1.1 Basic Concept of Entropy by Rudolf Clausius

The formulation of entropy as a fundamental physical quantity is attributed to the physicist Rudolf Clausius (Fig. 1.1), who intensively investigated the basics of thermodynamics.[1] Rudolf Clausius was born in Köslin (Koszalin) in 1822.[2] He came from a very large family. Despite having 17 siblings, he was able to attend the Gymnasium in Stettin (Szczecin) and study physics and mathematics in Berlin from 1840. Initially, he worked as a high school teacher and earned his doctorate in 1848 with a dissertation on the theory of the rainbow. Shortly thereafter, thermodynamics became the focus of his interest. As early as 1850, he discovered that the conversion of heat into work depended solely on the temperatures of the substances that absorbed or released the heat, and not on specific material parameters. His engagement with the direction of the process of cyclical operations

[1]R. Clausius, 1850.

[2]Biographical data of R. Clausius: W. Stiller, 1989, p. 84.

Entropy and Sustainability: Challenges of Future Energy and Material Management from a Thermodynamic Point of View
Michael Köhler

ISBN 978-981-5129-99-1 (Hardcover), 978-1-003-74889-2 (eBook)
www.jennystanford.com

DOI: 10.1201/9781003748892-2

eventually led him in 1865 to formulate the second law of thermodynamics and introduce a new central thermodynamic quantity, entropy.[3]

Figure 1.1 Rudolf Clausius (around 1850).[4]

Clausius had discovered that heat energy could not be raised from a lower to a higher energy level without additional effort, meaning that the direction of a spontaneous heat transfer is fixed, making the heat transfer process asymmetrical concerning the time axis. His new quantity, entropy, allowed for a quantitative description of this asymmetry. Clausius used this quantity to formulate the second fundamental law of thermodynamics: "The entropy of a system can only decrease if the entropy of another system increases by at least the same amount." From this fact, it immediately follows that in a closed system, entropy can increase but never decrease.

Clausius recognized the universal significance of his discovery. He understood that his observation of the direction of heat transfer and the introduction of entropy were expressions of irreversible

---

[3]R. Clausius, 1865.

[4]Figure taken from: commons.wikimedia.org/wiki/File:Clausius-1.jpg (last downloaded: 2025-01-10).

processes, and that natural processes, therefore, have a sense of direction (Fig. 1.2). This insight culminated in the application to a time arrow for the development of the cosmos as a whole: "The entropy of the world tends towards a maximum."

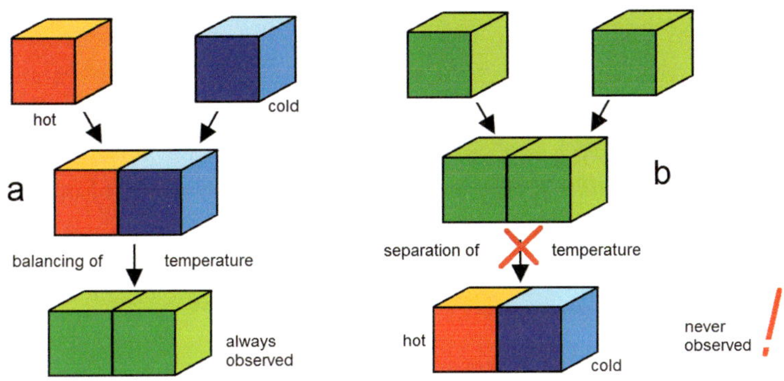

Figure 1.2 Irreversibility in the transfer of heat: Heat moves spontaneously from a hot to a cold body (left), but a spontaneous transfer of heat from a cold to a warmer body can never be observed (right).

The formulation of the second law of thermodynamics and the description of entropy were closely related to Clausius' work on the movement of particles in gas space. Clausius was a staunch supporter of the atomic theory, which was by no means undisputed among experts at the time. He saw that with the particle model of matter, its thermodynamic properties could be traced back to elementary mechanical laws. His engagement with this topic also gave rise to considerations of the vapor pressure of liquids and the investigation of the temperature dependence of vapor pressure, which eventually led to the formulation of the Clausius-Clapeyron equation.

On November 7, 1870, Rudolf Clausius presented his fundamental considerations on the mechanical foundations of gas theory to the Dutch Society for Natural and Medicinal Sciences and published them in 1871 as a written contribution in Poggendorff's Annalen. At that time, he was unaware of the young Ludwig Boltzmann's 1866 explanation of the second law of thermodynamics based on the principles of analytical mechanics. Both physicists independently arrived at this connection, and

Rudolf Clausius acknowledged Boltzmann's priority, which Boltzmann claimed after his publication.

The basis for Clausius' concept was the linkage of atomistic thinking with thermodynamics, specifically with the quantitative analysis of heat transfer (Fig. 1.3). This corresponded to the consistent linking of the macrocosm and the microcosm. At that time, in the last decades of the 19th century, the ideas about the atomic structure of substances were still quite hypothetical since neither direct experimental analyzes of the atomic structure of matter nor microscopic images of the building blocks of matter, as we use today, were possible.

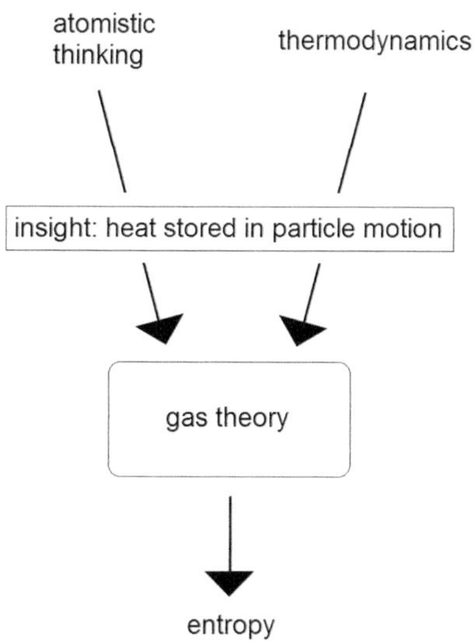

Figure 1.3 Conceptional connection between atomistic thinking and analytical thermodynamics allowed to formulate a theory of gases and to discover the fundamental role of the physical quantity named entropy.

The connection of the microcosm and the macrocosm in Rudolf Clausius' thinking also meant the clear idea of the universal and scale-independent validity of fundamental physical laws. Only with this conviction could the empirically obtained insights in the macroscopic world be transferred to the atomistic model of the microcosm. At the same time, recognizing the universality of physical laws also implied the understanding that processes in the microcosm determine the process flow in the macrocosm, that macroscopic measurable quantities such as heat, temperature, and entropy are expressions of the behavior of large ensembles of small particles. With the mechanical justification of the second law, it became clear that entropy is not only a central thermodynamic quantity but can also be derived from the laws of statistics. The way the components of matter interact is governed by the laws of mathematics.[5]

1.2 Loss of Structure in Spontaneous Processes

The decline and formation of structures are everyday phenomena that can be observed in many processes in industry and daily life. The creation and disappearance of structures should not be understood as mirror-image processes. To illustrate, a simple example of the disappearance of a structure in a technical process is described below. Large numbers of small grains of limestone, soda, and small quartz crystals are mixed and heated. The crystals touch each other, and where the material comes into contact, it begins to stick and then melt. Bubbles rise, and the solid grains become components of a slushy, cloudy mass. Gradually, this viscous mass becomes clearer and more mobile. Finally, the cores of the quartz crystals dissolve in the now-formed liquid. From small solid particles of silicon dioxide, soda, and calcium carbonate, a glass melt is formed. Glassmakers take this melt, often with additional additives, and shape it into forms and functions: windows, household items, glass tubes, ampoules, laboratory glass, and much more. The heating and melting of the raw materials enable this manufacturing process and the creation of useful and aesthetic structures. However, for such shaping to be possible, existing structures must first disappear. The previously separated raw materials have dispersed on an atomic level in the glass matrix. The atoms did this spontaneously after being brought together and heated enough to gain sufficient mobility. After melting, this spontaneous movement, known as Brownian motion, alone caused the disappearance of the spatial arrangement in the

[5]G. Ronge, 1956.

previously existing crystals, leading to a mixing of the previously separated silicon, calcium, and sodium atoms. The transition from the regular arrangement of atoms in the crystal lattices to the largely random arrangement of atoms in amorphous glass is the main driving force of this process. It corresponds to the loss of order in the crystal lattices and can be characterized by an increase in entropy (Fig. 1.4).

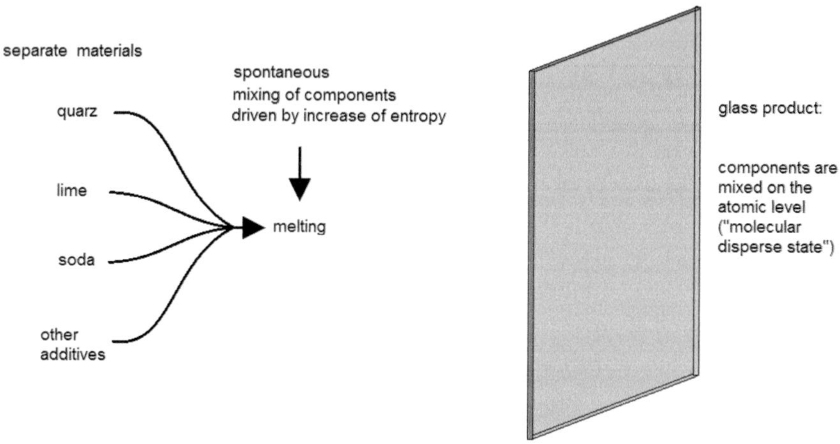

Figure 1.4 Destruction of the crystal order of separate material components during the formation of the amorphous structure of glass.

At low temperatures, when the crystals are stably adjacent, a state of lower entropy prevails. After melting and possibly subsequent solidification, a state of higher entropy exists in the glass melt or glass. The described material system, when melted, follows, as expected, the Second Law of Thermodynamics, which can be best rediscovered in the following formulation: In a closed system, entropy never decreases. Spontaneous changes in the system are inevitably linked to an increase in entropy.[6]

Human activities, both technical and natural, are practically always tied to the breakdown of structures at the level of the elementary building blocks of matter, atoms, and molecules. To build and maintain macroscopic structures, microscopic structures must be destroyed. Examples of this can be found

[6]P. Atkins, 1984.

everywhere in daily life. Crystal order, for example, is broken down, and substances are distributed when salt is dissolved in potato water or sugar in tea. For a tire to roll with good traction on the road, rubber material in the form of hydrocarbon fragments is transferred from the tire surface to the road. For dishwashing, handwashing, or laundry washing, soap is dissolved and released in the form of a soap solution into the sewer system, or, in the less favorable case, directly into the environment. Our nutrition is also such a process. The energy stored in plants in the form of carbohydrates, fats, and proteins is needed by us for life. The destruction of the plant as a structure and the breakdown of its components down to small molecules like water and carbon dioxide is a vital process. Through our breathing, the carbon that was previously built into leaves, roots, and fruits in the form of organic compounds is released as carbon dioxide and distributed in the atmosphere. Humans, along with all other living beings, are producers of entropy (Fig. 1.5).

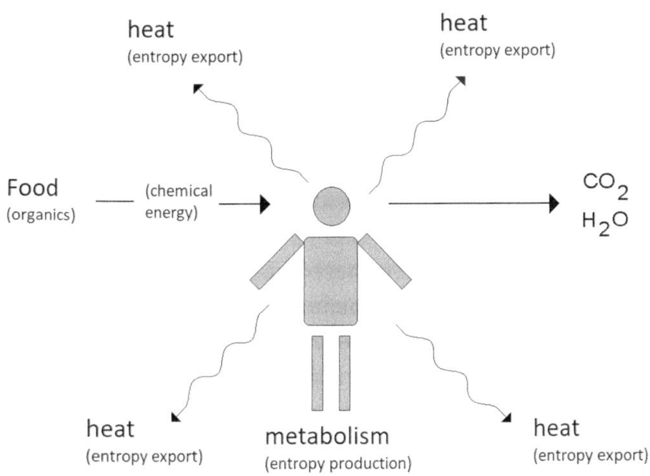

Figure 1.5 Human beings produce entropy by their metabolism and export the entropy to the environment by the release of heat and low-energy small molecules.

Humans and animals build their structures by utilizing molecular building blocks and obtaining them from the structures

Similarly, k_B can also be derived from the thermal energy in connection with the Nernst equation, which describes the concentration dependence of the electrochemical potential (here for the example of an electrochemical redox reaction):

$$k_B = (U - U_0) \times z \times q_e / [T \times \ln (a_{ox}/a_{red})];$$
$$\text{derived from: } U = U_0 + [k_B \times T/(z \times q_e)] \times \ln (a_{ox}/a_{red}). \quad (1.8)$$

U is the concentration-dependent electrochemical potential, U_0 is the standard potential characteristic of a process, z is the electrochemical valence, i.e., the number of charges transferred per elementary step, q_e is the elementary charge, and a_{ox} and a_{red} are the activities, i.e., the concentrations corrected by activity coefficients, of the species involved in potential formation in their oxidized and reduced forms.

In all these examples mentioned above, the statistics of the movement of atoms and molecules are expressed. The Boltzmann constant appears as a fundamental quantity with the dimension of heat capacity. Entropy also has the dimension of heat capacity. In this sense, the Boltzmann constant can be regarded as a fundamental entropy. Its central role arises as a direct consequence of the discrete, i.e., atomic or molecular, structure of matter. It describes the energetic relationships in the momentum transfer and movement of the building blocks of matter, and thus, in a way, the lowest level at which impulses and kinetic energy can be distributed. This momentum transfer in the random movements of particles is the final stage of the degradation of mechanical energy, which we perceive macroscopically as heat.

1.7 Process and Reaction Rates

The quantitative description of the course of chemical reactions is the subject of chemical kinetics. Typically, processes are considered in which vast amounts of identical particles participate in a reaction. The temporal change in the concentrations of reactants and reaction products is determined by statistical laws. The reaction rates are higher, the greater the concentrations of the involved particles and the greater the likelihood that a reaction will occur upon the encounter of particles. This is easiest to understand with the example of a so-called monomolecular process. In this case, the probability of the process itself does not depend on the reaction partners and is solely determined by a characteristic constant value, the rate constant k. Therefore, the relative reaction rate, meaning the rate related to the concentration of the reactants, remains constant. However, the absolute reaction rate decreases with the diminishing concentration of the reactants, with the initially high reaction rate decreasing more quickly at first and then slowing down. This is described by an exponential function, which applies to this class of chemical reactions as well as to the decay of radioactive substances:

$$C_t = c_0 \times e^{-k \times t}. \tag{1.9}$$

The rate constant k corresponds to a characteristic process time τ:

$$\tau = 1/k. \tag{1.10}$$

Therefore, this equation can also be written in the following form:

$$t/\tau = \ln [c_0/c_t]. \tag{1.11}$$

This equation represents an important example of processes characterized by a statistical relationship—in this case, the concentrations. The ratio of the concentrations arises from the initial concentration and the probability with which reactions occur within a specific time interval. This determines the concentration that will be reached after a certain period of time. The concentration ratio $[c_0/c_t]$ represents the reaction probability in the corresponding time interval. The initial concentration at time t = 0 can be regarded as a reference quantity. The ratio of this reference quantity to the changing concentration $[c_0/c_t]$ is exponentially dependent on the process variable time, where time t also relates to a reference quantity—the characteristic process time τ.

In any case, a directed process is described. A system with a high concentration of the starting material transitions to a state where a lower concentration of that substance exists. The temporal progression of the relative concentration change is uniquely determined by the characteristic process time, independent of the specific time.

This chemical process is driven by the chemical potential, a form of free energy, *F*. During the reaction, the free energy is reduced, and entropy is produced as the system approaches the equilibrium state.

1.8 Svante Arrhenius

Around the turn of the 18th to the 19th century, the determination of the masses of reactants and products led scientists such as Lavoisier, Proust, and Dalton to establish the fundamental laws of stoichiometry and thus to an empirically grounded atomic hypothesis. It was ultimately these mass determinations that, a good half-century later, provided Dmitri Mendeleev and Lothar Meyer with the key to formulating the periodic table of chemical elements. However, this did not dispel all doubts about the atomic structure of matter. The natural sciences of the second half of the 19th century remained defined by the debate surrounding the corpuscular nature of substances. The broader application of physical measurement methods to substances and their reactions led to a turnaround toward the end of the 19th century. With the development of physical chemistry, insights were rapidly gained that fit well into the picture of matter composed of discrete particles. Alongside physicists like Rudolf Clausius and Ludwig Boltzmann, it was primarily scientists working in physical chemistry who contributed to the swift advances in understanding the properties of substances. Among them, Svante Arrhenius was a particularly outstanding and influential figure in the field.

Svante Arrhenius (Fig. 1.12) was born on February 19, 1859, at the Vik estate near Uppsala. He attracted attention as a child due to his high ability, studied quickly and very successfully in

Uppsala, and the diverse range of subjects he chose—including mathematics, chemistry, and mineralogy—became important for the direction of his scientific work. Right at the beginning of his research in Stockholm, the phenomenon of dissociation was at the forefront, and by the age of 24, he was able to lay down the fundamentals of his theory of dissociation. He submitted his findings on the behavior of electrolytes for his doctorate at Uppsala University in 1884 but could only achieve the modest distinction of "non-sine laude." Only the chemist Otto Pettersson recognized the outstanding importance of the work in Stockholm. However, Arrhenius was undeterred by the disappointing response and sent his dissertation to leading international scientists, including Rudolf Clausius in Bonn, Lothar Meyer in Tübingen, Wilhelm Ostwald in Riga, and van't Hoff in Amsterdam. Wilhelm Ostwald was particularly fascinated by the work, wrote back, and traveled to Uppsala in August 1884 to meet the young Arrhenius. This trip marked the beginning of a lifelong and scientifically extraordinarily fruitful friendship.[15]

Figure 1.12 Svante Arrhenius.[16]

[15]Biographical data of S. Arrhenius: E.H. Riesenfeld, 1931.

[16]Figure taken from: commons.wikimedia.org/wiki/File:Svante_Arrhenius.jpg (last downloaded: 2025-01-10).

In 1886, Svante Arrhenius reciprocated Ostwald's visit with a working stay in Riga. In the winter of 1886/87, he experimented at Kohlrausch's institute in Würzburg, where he coincidentally met Walther Nernst, who was working there at the same time. It was Nernst who persuaded Svante Arrhenius to join him in Graz for the summer semester of 1887. There, Ludwig Boltzmann held the chair in physics and was an important attraction for young scientists. This group also included Wilhelm Ostwald, who inspired the initially physics-focused Walther Nernst to engage with physical-chemical problems. The months in Graz, Boltzmann's charisma, and the lively exchange among the younger scientists invigorated the development of physical chemistry as a new, independent scientific field.

After Svante Arrhenius had worked in his native Sweden over the winter, he visited Max Planck in Kiel in spring 1888 and then van't Hoff in Amsterdam, where they discussed electrolytic dissociation, the law of mass action, and its application to strong and weak electrolytes. Subsequently, he spent the rest of the summer semester in Leipzig at Wilhelm Ostwald's laboratory, where Walther Nernst also worked and was just completing his work on the electromotive effect of ions. During this time, Svante Arrhenius focused on the relationship between osmotic pressure and vapor pressure while continuing his studies on dissociation. One result of this work was the paper "On the Heat of Dissociation and the Influence of Temperature on the Degree of Dissociation of Electrolytes."[17] The proof of a universal neutralization heat during the reaction of a strong acid with a strong base presented in this paper was a cornerstone in support of the theory of dissociation. Arrhenius's engagement with the temperature dependence of chemical reactions also led him to discover the role of activation energy. He reported on this in 1889 in the publication "On the Reaction Rate in the Inversion of Sucrose by Acids." With the equation $k = \text{const} \times e^{-Q/RT}$, he laid the foundation for describing the rate of chemical reactions as a function of the activation barrier and temperature.[18]

---

[17] S. Arrhenius, 1889a.

[18] S. Arrhenius, 1889b.

While Svante Arrhenius had already received great recognition as a young man among the circle of international physical chemists, his professor colleagues in Sweden regarded him with skepticism. Consequently, Arrhenius was always happy to accept invitations to work abroad, such as to Würzburg and Graz. He especially found a warm welcome in the laboratory of his supporter and friend Wilhelm Ostwald in Leipzig. After being appointed to a newly established extraordinary professorship in Gießen in 1891, Arrhenius finally received an offer for a lectureship in physics in Stockholm, which he gladly accepted. Although he greatly appreciated traveling to other countries and laboratories during lecture breaks and the direct contact and exchange with his foreign colleagues, he remained connected and loyal to his Swedish homeland throughout his life, despite all the skepticism and rejection he often faced at home.

1.9 Arrhenius Equation

The Arrhenius equation represents an empirical relationship between the rate constant k of a chemical reaction and the activation energy E_a of that reaction, more precisely, the ratio of the activation energy to the thermal energy $E_{th} = k_B T$ (Eq. (1.7)):

$$k = f \times e^{(-E_a/E_{th})}. \qquad (1.12)$$

The term f is a frequency factor typical for the process, which indicates how often a movement occurs that could potentially lead to a reaction if sufficiently high kinetic energy were present. The exponential factor indicates the proportion of particles that have energy above the activation threshold E_a, i.e., have enough kinetic energy for the reaction to proceed.

It is easy to see that this equation has the same structure as the equation describing the concentration course of a first-order reaction (see above). If, instead of k, the characteristic process time is used and instead of a characteristic movement time ($\tau_0 = 1/f$) is used, it follows that:

$$1/\tau = (1/\tau_0) \times e\,(-E_a/E_{th}) \qquad (1.13)$$

or

$$E_a/E_{th} = \ln(\tau/\tau_0). \tag{1.14}$$

This means that the activation energy is numerically related to the thermal energy in the same way as the logarithm of the ratio of the probability that a reaction occurs to the probability of the particle movement that could lead to the reaction (Fig. 1.13).

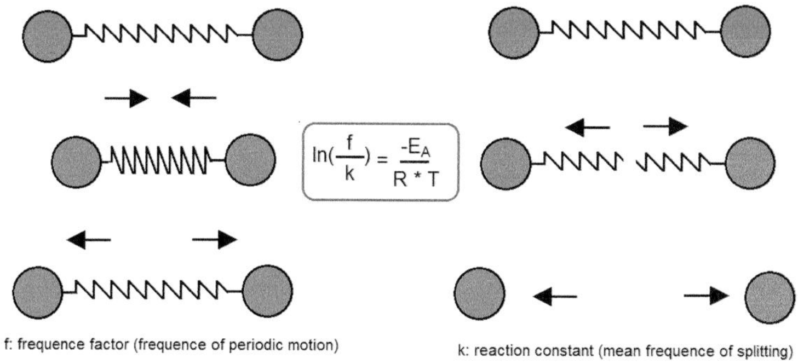

Figure 1.13 Probabilistic illustration of Arrhenius' law for a first-order reaction (homolytic bond cleavage): comparison of frequency of molecular motion (oscillation) in the direction of molecular splitting process (left) and probability of really occurring bond cleavage (right).

1.10 Critical View on Energy and Atmosphere by Svante Arrhenius

Undoubtedly, Svante Arrhenius made significant contributions to the establishment of physical chemistry as an independent science and developed his true area of expertise through a multitude of thorough experimental and theoretical investigations. At the same time, his broad scientific interest enabled him to achieve important and partially revolutionary insights in other fields of science, where the modern understanding of physical-chemical problems was always helpful. A good example of this is his hypothesis concerning the formation of potassium salt deposits. He interpreted the elevated temperatures required for the formation of potash by a sequence of sedimentation, overlaying by younger sediments up to 2 km in height, and subsequent erosion of the overlying layers. These thoughts, rooted in the idea of long-term directed changes in geological development, initially faced massive opposition from experts, which Arrhenius countered with a clever argumentation during the Kalitag in Göttingen in 1912.[19]

In addition to physical chemistry, mineralogy, and cosmology, Arrhenius was also interested in meteorology. As early as 1888, he studied phenomena of atmospheric electricity and identified shortwave solar radiation as the cause, which leads to ionization in the atmosphere. In 1897, he reported in Stockholm on his ideas about the cause of the Northern Lights, which he correctly explained as the result of ion influx into the Earth's atmosphere, clearly noting the connection between solar activity and the auroras. He was also able to explain the formation of comet tails through this stream of particles coming from the Sun. The effect of the solar wind on other celestial bodies and the transport of particles through the cosmos reinforced his assumption that primitive forms of life, such as spores, could have been transported through the cosmos and thus disseminated. His formulated panspermia hypothesis was often met with skepticism, but fundamentally, it now serves as a working hypothesis for the current search for extraterrestrial life. His universal scientific thinking is also reflected in his cosmological and popular science works, including the "Textbook of Cosmic Physics" (1903) and the extraordinarily popular and widely translated book "The Origin of Worlds" (1906) and "Concepts of the Universe through the Ages" (1908), as well as a whole range of scientifically and popular science-oriented articles on astrophysics.[20]

Shortly after the turn of the century, Svante Arrhenius concerned himself with the issue of the Earth's energy supply. He pondered how long humanity would be able to rely on the reserves of oil and coal from which industrial society was drawing. The early timing of his engagement with this question deserves respect today, considering that motor vehicles had only recently been invented, horse-drawn vehicles were still the primary means

[19]E.H. Riesenfeld, 1931, 73.

[20]E.H. Riesenfeld, 1931, 63 and following pages.

of transportation alongside the railway, and there was no aviation yet. In a remarkable foresight, he predicted that solar power plants in the lower latitudes and wind power plants in the middle latitudes would become the main energy sources for humanity (Fig. 1.14).

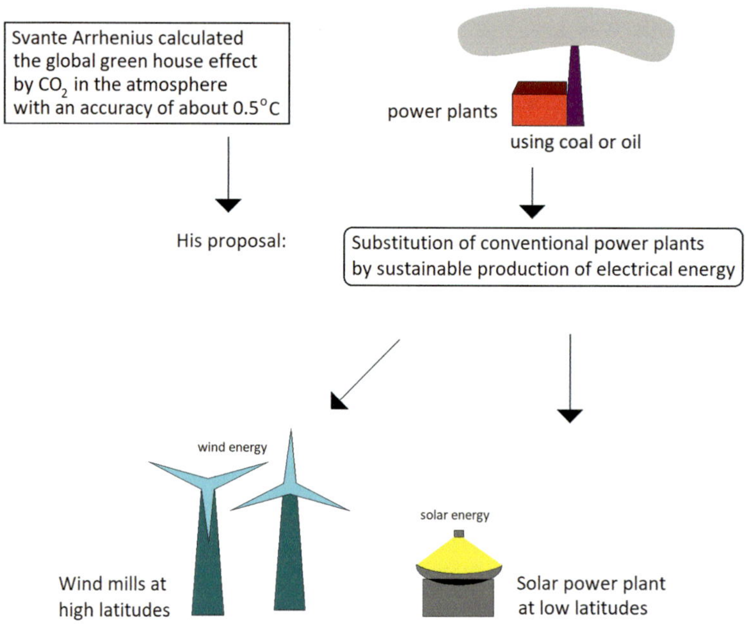

Figure 1.14 Arrhenius' brilliant analysis of the CO_2-caused greenhouse effect and his early insight into the need to substitute the burning of coal and oil by sustainable energy production.

These considerations are brilliantly confirmed by the current ongoing developments. It is particularly noteworthy that Arrhenius was guided by the concept of sustainability in energy supply even then, recognizing the necessity of linking humanity's energy supply to so-called renewable energy sources, i.e., he understood that industrial energy supply must be coupled with the energy flow from the hot sun through the Earth to the cold cosmic background.

Equally groundbreaking was Arrhenius' insight into the role of atmospheric carbon dioxide in climate development. In 1896, his work "On the Influence of the Atmospheric Carbon Dioxide Content on the Temperature of the Earth's Surface" was published.[21] He recognized the greenhouse effect of carbon dioxide due to its infrared absorption and interpreted the geologically evidenced shifts in vegetation belts and temperatures during ice age cycles in relation to changes in the CO_2 content of the atmosphere, estimating that the necessary fluctuations amounted to about 50% of the current concentration (Fig. 1.15). In response to the objections from the scientific community, he addressed them with a reassessment of the infrared absorption by CO_2 in Berlin in 1900, although he did not achieve full recognition of his theory at that time.[22] Modern measurements from ice cores and climate models have since splendidly confirmed Arrhenius' theory.

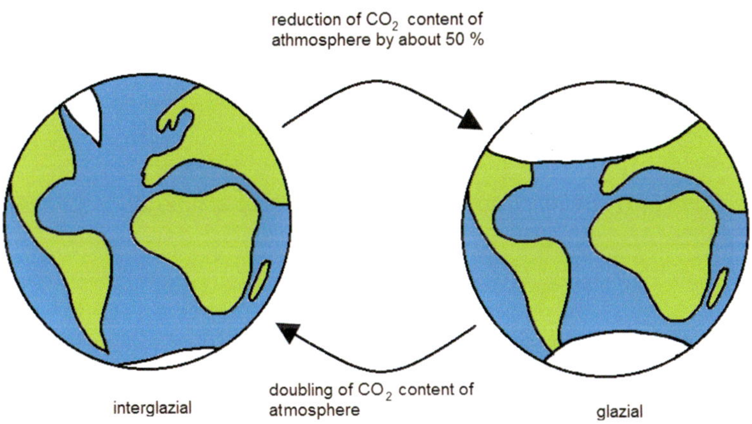

Figure 1.15 Estimation of oscillating strength of greenhouse effect between glacial and interstadial phases of ice age due to oscillating global CO_2 concentration in the Earth's atmosphere by Svante Arrhenius.

1.11 Thermodynamic Aspect of Information Management

The acquisition and processing of information also require energy. This truism is already taught to us by the simple fact of

[21] S. Arrhenius, 1896.

[22] E.H. Riesenfeld, 1931, 41.

the power consumption that computers have. A mechanical calculator receives its energy in a mechanical way. In both cases, information can be accumulated or generated. Energy is consumed by the system but is not accumulated, only converted. Like most living systems, the mechanical calculator and the computer can be understood as energetically balanced systems. This means that the incoming mechanical or electrical power corresponds exactly to the outgoing power, which is released as heat to the environment (Fig. 1.6). From an energetic point of view, the calculating machines are stationary—the machine releases as much energy as it absorbs.

Energy conversion is inevitable when information needs to be processed. But how large is the minimum energy required to obtain a certain amount of information? The thermodynamic limit is described by the Boltzmann equation. For the minimum information content of 1 bit, it is given by:

$$E/E_{th} = [E/(k_B \times T)] \times \ln(2). \tag{1.12}$$

For any given information with a size of z bits, the following applies:

$$E/E_{th} = \ln(z). \tag{1.13}$$

The quotient E/T can be understood as entropy when there is uncertainty about the occurrence of exactly one of z possibilities:

$$E/T = S = k_B \times \ln(z). \tag{1.14}$$

For this entropy, the Boltzmann constant serves as a reference value, just as thermal energy serves as a reference value for the corresponding energy:

$$S/k_B = \ln(z). \tag{1.15}$$

At a room temperature of approximately 300 K, this entropy of 1.38×10^{-23} J/K corresponds to an energy of about 4×10^{-21} J per bit. The Boltzmann equation indicates that the energy consumed in information processing cannot be reduced arbitrarily.

The Boltzmann constant provides the lower limit for these operations. A smaller value is not possible. The absorption of energy, such as in the form of electrical energy, and the release of energy as heat are causally related. Both are essential for the information processing process. The flow of energy through the system—“energy dissipation“—is unavoidable. The heat output of a computer is just as essential as the intake of electrical energy. The two processes are two sides of the same coin.

processes and irreversible changes. Directed processes include cosmic expansion and the global cooling of the universe (Fig. 2.2).

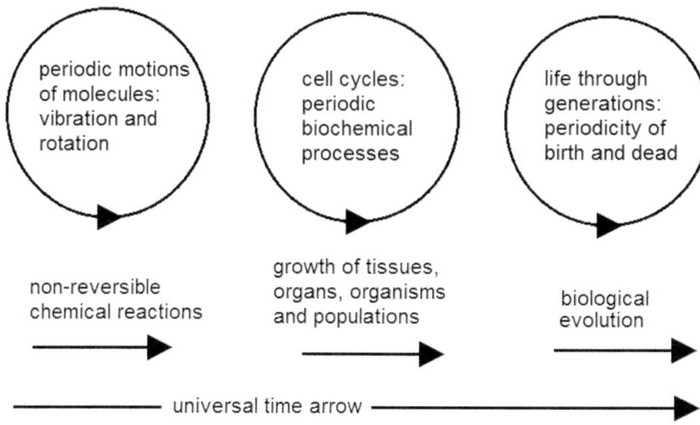

Figure 2.1 Superposition of cyclic and non-reversible processes as a general phenomenon in nature.

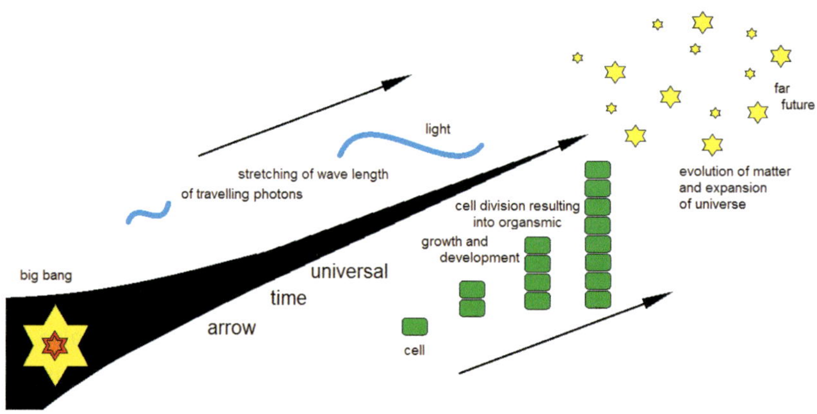

Figure 2.2 Time arrow in cosmic evolution.

Even the single photons, which are travelling through the universe for millions or billions of years, experience irreversibility. It manifests in the "aging," i.e., the reduction of photon energy due to the expansion of space. Directed processes control large structures in the universe, too. The enlargement of gigantic voids between galactic superclusters and the dwindling dynamics of aging galaxies are irreversible. Irreversibility determines the evolution of matter, the formation of heavy elements through nucleosynthesis in stars (Fig. 2.3).[2]

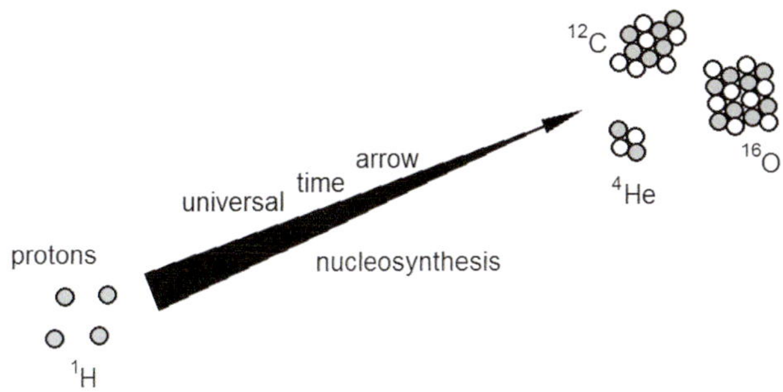

Figure 2.3 Irreversibility in the formation of matter in the universe by nucleosynthesis in stars.

The formation and development of stars is a non-reversible far-equilibrium process. Irreversibility is evident in the aging of remnant stars, in the succession of thermonuclear processes in the life of larger stars, in their collapses, explosions and formation of bloated giant stars (Fig. 2.4). In case of smaller stars, their final collapse into highly dense, so-called degenerated matter in white dwarfs is a non-reversible process, too (Fig. 2.5).

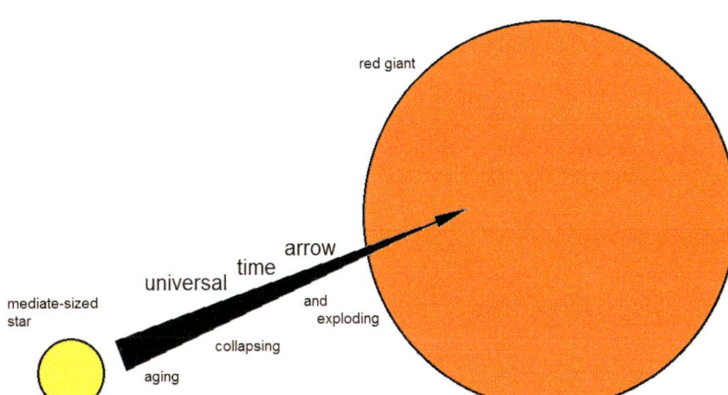

Figure 2.4 Irreversibility in critical phases of stars: collapse and bloating of stars during the transition between different development phases and the start of new thermonuclear reactions.

[2]P.C.W. Davis, 1977.

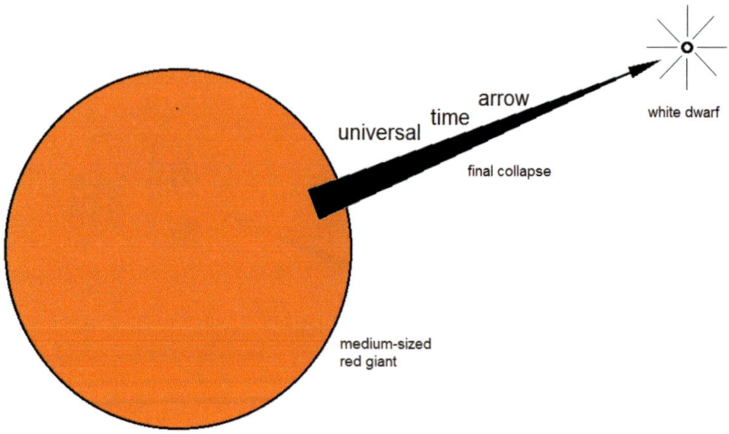

Figure 2.5 Irreversibility in the collapse of small- and medium-sized stars into White Dwarfs.

And so is the fate of heavy stars which collapses finally into neutron stars, and black holes accompanied by the incredible intensive explosion in form of supernovae (Fig. 2.6). In all these processes—from cosmic dimensions to microscopic, biomolecular processes, in both non-living and living nature—directed processes dominate in the long term, like an unyielding indicator determining the path that systems must follow. This is the arrow of time.[3]

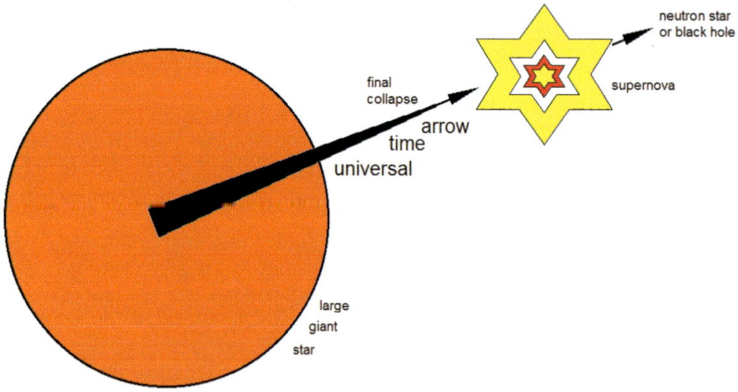

Figure 2.6 Time arrow in the development and collapse of large stars in supernova outbreaks and formation of neutron stars and black holes.

[3]S. Hawking, 1988.

At first glance, cyclic processes seem to reproduce initial states repeatedly. However, if one considers these processes on a sufficiently large timescale, it becomes evident that there is no absolutely precise repetition, and the initial state is not entirely restored. Even behind cyclic processes, the arrow of time is always present. Cyclic processes are nothing more than more or less pronounced modulations of this arrow of time. The arrow of time, i.e., the fundamental irreversibility of all events, is probably the most important general law of our universe. The examples illustrate the character of the arrow of time. It is evident that our everyday experience is also determined by the arrow of time.[4]

Where do we encounter the arrow of time in daily life? Why does our world have the character of a one-way street? What is the basis of the arrow of time?

2.2 Second Fundamental Law of Thermodynamics

At the beginning of this section, we will consider simple thought experiments in which we assume a balanced energy budget but imagine that processes are reversible. These thought experiments are meant to highlight how accustomed we are as humans to the effects of the arrow of time. First, let us consider a hypothetical simple cart. If it starts from an elevated position of an inclined plane, it starts to move. Potential energy is converted into kinetic energy. This energy is further converted into heat by friction processes. As a result, the car will stop after rolling into a plane level (Fig. 2.7). The total of all involved forms of energy (potential energy, kinetic energy, and thermal energy) remains constant. The first fundamental law of thermodynamics is fulfilled. All observations correspond well with our everyday experience.

In the second thought experiment, we regard the same simple cart. Now, it is in the flat plane. But, by uptake of heat from the environment, it starts to move. By further conversion of heat energy into kinetic energy, the cart is accelerated. Thus, it can roll upwards. Finally, it comes to an elevated position. In the whole

[4]I. Prigogine, 1992.

process, heat energy is converted into potential energy. The total of all involved forms of energy (potential energy, kinetic energy, thermal energy) remains constant, corresponding to the validity of the first fundamental law of thermodynamics in this case (Fig. 2.8).

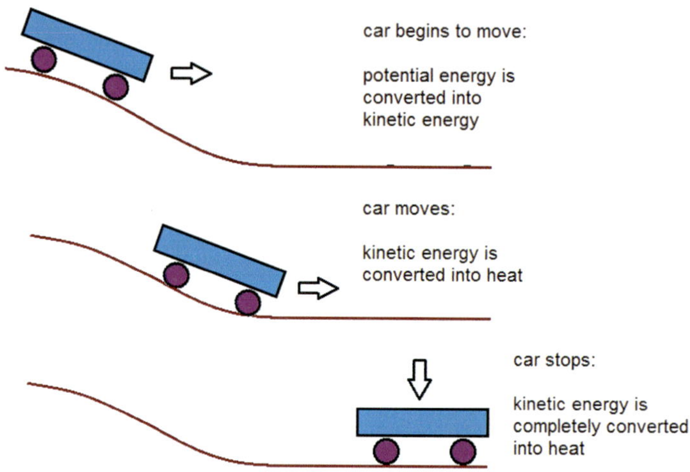

Figure 2.7 Energy conversion in the case of a rolling cart starting from an elevated position.

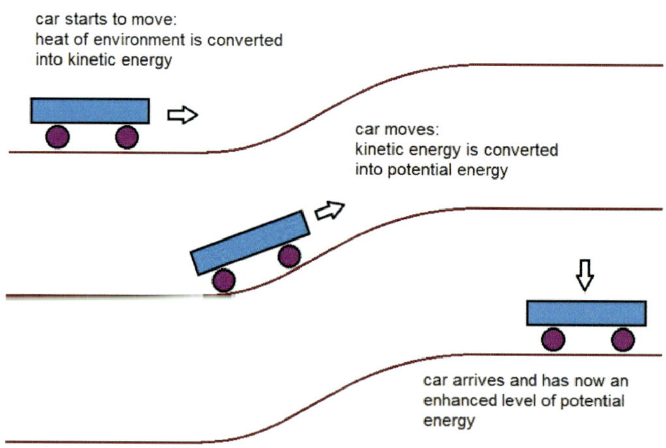

Figure 2.8 Energy conversion in the case of a hypothetical rolling cart starting from a low position and arriving at an elevated position after using heat from the environment for the generation of kinetic and potential energy.

The law of conservation of energy is completely fulfilled as it was in the previous example. However, the assumed direction of energy conversion contradicts our everyday experience. Something is wrong.

In a further example, we regard an automobile of a special kind. We assume that all processes occurring in the car are reversible. A normal car derives its energy for driving from the combustion of fuel. In this process, hydrocarbons, as found in gasoline or diesel oil, react with oxygen from the air to produce carbon dioxide and water. In our hypothetical car, the engine draws carbon dioxide and water from the atmosphere. The engine converts these substances into liquid hydrocarbons. The energy required for this is obtained from cooling the ambient air via the radiator. As a byproduct of the chemical conversion, oxygen is released into the atmosphere. The energy extracted from cooling the air is greater than the energy needed for the chemical reaction. The difference is converted into mechanical energy, which provides the car with its mobility. The resulting hydrocarbons are collected and occasionally pumped out at a disposal station, usually for a substantial credit.

This example seems absurd at first glance. It completely contradicts everyday experience, too. Yet, from the perspective of the law of conservation of energy, it is entirely correct. The vehicle is not a perpetuum mobile. The energy balance is intact. The error lies in the causality of the processes. No machine can perform mechanical work through the spontaneous cooling of its surroundings. Nor can energy-rich chemical compounds, electrical energy, or light be generated this way. Such hypothetical machines are referred to as "perpetuum mobile of the second kind." Like the perpetuum mobile that violates the First Law of Thermodynamics, this too cannot exist. The perpetuum mobile of the second kind contradicts the Second Law of Thermodynamics. Its process would run counter to the arrow of time.

The Second Law of Thermodynamics applies not only to heat transfer and chemical processes but also governs mechanical operations and the processing of information.[5] These, too,

---

[5]P. Atkins, 1984.

sense, the transition from a more ordered to a less ordered state. The ordered state is characterized by a spatial separation of the thermally more excited particles from the less excited particles. After thermal equilibrium is reached, the less ordered state exists, in which all particles are thermally excited in the same way on average.

The transition from the ordered to the less ordered state occurs spontaneously. It is always associated with a reduction in free energy. This reduction in free energy is the driving force of the process, which is why it is irreversible. The arrow of time underlies this process.

Apart from the dynamics in the microcosm, such as momentum transfer between the particles of a gas or in chemical equilibrium, nearly all processes are irreversible. Macroscopic processes involving bodies composed of multiple molecules and atoms always occur with a loss of free energy. This is simply because systemic order and/or systemic energy is always transferred from the macroscopic level to the microscopic level. This is most evident in the case of mechanical motion: Macroscopic movements, such as driving a car or a ship, require overcoming frictional forces, air, or water resistance. In all cases, mechanical energy is converted into frictional heat. This conversion is nothing more than the transfer of part of the macroscopic momentum to the movement of particles, which manifests as an increase in heat content and thus a measurable increase in temperature. Quantitatively, the energy ΔQ transferred in this process can be described as an increase in entropy:

$$\Delta S = \Delta Q/T. \tag{2.1}$$

The second fundamental law of thermodynamics says that in thermodynamically closed systems, the entropy is constant or must increase. It is constant if the system is in thermodynamic equilibrium. Besides microscopic fluctuations, no processes go on (Fig. 2.11, left). The entropy increases if the closed system is changing (Fig. 2.11, right). Spontaneous processes can only go on if the system is not in equilibrium. Each increase of entropy relates to an approach to the equilibrium state.

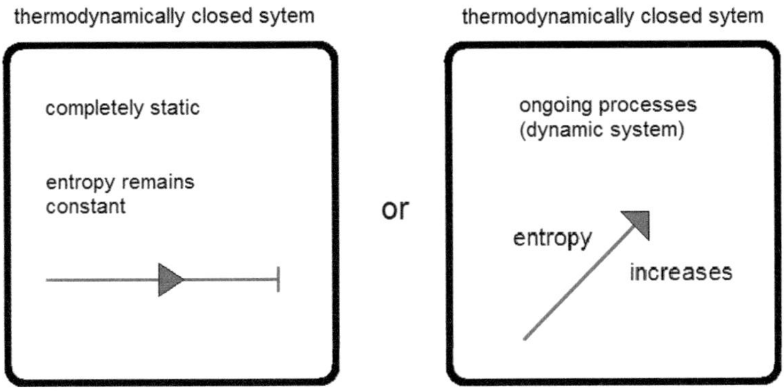

Figure 2.11 Schematic illustration of the second fundamental law of thermodynamics. In a closed system, the entropy is constant or can increase; a decrease in entropy in a closed system is not possible.

When energy is transferred from the macroscopic to the microscopic level, the level of atoms and molecules represents a fundamental limit. Once thermal equilibrium is established, entropy can no longer increase. Figuratively speaking, there are no smaller particles to be disordered than atoms and molecules. In fact, this picture holds only as long as the particles themselves do not change. Chemical structures within the molecules can be an additional source of free energy beyond the transfer of mechanical impulses, and their breakdown would lead to further irreversible processes.

The increase in entropy at the level of atoms and molecules due to the equalization of temperature between two bodies is determined by their heat capacity. Entropy itself has the dimension of heat capacity, and by definition, it is determined by the ratio of exchanged heat to absolute temperature. For microscopic processes, a characteristic energy can be given for each temperature, which has the character of an elementary energy related to a particle or a degree of freedom of movement. The ratio of this energy to absolute temperature represents an elementary entropy. This quantity is the Boltzmann constant k_B.

The statistical formulation of entropy (see above) can be easily rewritten so that the order of a system is represented as the ratio of its entropy to this reference quantity:

$$S/k_B = \ln(w_0). \tag{2.2}$$

Here, w_0 denotes the number of microstates that represent the system. If, as in the case of rolling dice or drawing lots, multiple outcomes are possible, their number w enters the equation:

$$S/k_B = \ln(w_0/w). \tag{2.3}$$

As the number of possible outcomes w increases, the system becomes more defined, which is equivalent to decreasing entropy. This principle applies not only to atoms and molecules but to all systems composed of components that can be ordered or disordered. Even on the macroscopic level, effort is required to create order. The Boltzmann equation can thus be generalized by replacing S with a quantity Y that represents a general effort and k_B with a reference measure for effort Y_0. On the right side of the equation, the statistical probability, a ratio of quantities, a ratio of time spent to reference time, or the corresponding frequency ratio, is represented:

$$Y/Y_0 = \ln(w_0/w). \tag{2.4}$$

As already seen above, other central equations of physical chemistry can also be expressed in this general form. A simple example is the kinetic time law for a first-order chemical reaction. The ratio of the time required t to reach a certain concentration c to the characteristic reaction time $\tau = 1/k$ is determined by the initial and target concentrations:

$$t/\tau = \ln(c_0/c). \tag{2.5}$$

The analogy becomes even clearer in the case of the Arrhenius equation (Eqs. (1.14) and (2.6)). The activation barrier E_a to be overcome represents an effort that must be made. The thermal energy E_{th} ($=k_B \times T$) is the corresponding reference quantity. The number of possibilities that could potentially lead to a reaction is expressed by the frequency factor f, while the frequency of events that result in a reaction is expressed by the rate constant k:

$$E_a/E_{th} = \ln(f/k). \tag{2.6}$$

The processes described by this equation are irreversible, meaning they are subject to the arrow of time. For processes at the microscopic level, the Boltzmann constant serves as the reference value as an elementary measure of entropy.

In general, microscopic processes, which can be statistically well-described by a large number of particles, can be captured by an equation of this form. The numerical proportion of active particles is well-determined by the Boltzmann factor, where the exponent contains the ratio of the activation barrier to thermal energy. This also applies to the formation of electrochemical potential (Eq. (1.8)), where, in the case of redox electrodes, the activity ratio of electrochemically effective oxidizing and reducing species takes the role of probability. The logarithm of this ratio equals the ratio of the characteristic electrochemical energy ΔE_{el} to the thermal energy E_{th}:

$$\Delta E_{el} / E_{th} = \ln(a_{ox}/a_{red}) \tag{2.7}$$

with $\Delta E_{el} = (E - E_0) \times z \times F$ and $E_{th} = k_B \times T$,

where E and E_0 represent the concentration-dependent electrode potential and the standard potential. This equation is simply an alternative expression of the Nernst equation for potential formation under equilibrium conditions at redox electrodes:

$$E - E_0 = [k_B \times T/(z \times F)] \times \ln(a_{ox}/a_{red}). \tag{2.8}$$

There is, of course, an analogous formulation for equilibrium in thermal reactions. Here, the standard free reaction enthalpy ΔG_0 takes the place of electrochemical energy, and the logarithm contains the equilibrium constant K as the ratio of the rate constants of the forward and reverse reactions k_1 and k_2:

$$\Delta G_0 / k_B T = \ln(k_2/k_1) \quad \text{or} \quad \Delta G_0 = - k_B \times T \times \ln(K). \tag{2.9}$$

However, the general form of equation (Eq. (2.4)) also applies to macroscopic systems. The only difference is that other reference values need to be considered. Probably, there exist a number of up-to-now undiscovered laws which can describe spontaneous processes, loss, and accumulation of information on higher, and more complex levels of hierarchically organized far-equilibrium systems.

The comparison between statistically and thermodynamically defined entropy can also be used for judging the energy consumption for operating computers. The theoretical limit for the energy consumption E_{min} for a computer operation of 1 bit is given by the Boltzmann equation and the operating temperature:

$$E_{min} = k_B \times T \times \ln(2). \tag{2.10}$$

At room temperature, this is approximately 3×10^{-21} J (as mentioned above). This energy is about equal to the electrical energy used to accelerate a single electron in a potential of 0.1 V. However, the actual energy required for the switching process in a computer is much higher and can be calculated based on the power consumed. For simplicity, if we assume the total consumed electrical power as the effort E, the reference effort E_0 can be calculated using the operating frequency for the elementary operation:

$$E_0 = E / \ln(2). \tag{2.11}$$

The following can be derived assuming a power consumption of 30 W per processor core and a bit-related clock rate of 3 GHz:

$$E_0 = 30\ \text{W} / (3 \times 10^9\ \text{s}^{-1} \times \ln(2)) = 1.4 \times 10^{-8}\ \text{J}. \tag{2.12}$$

This value is the reference value for the computer's energy efficiency. It can be considered a system constant that allows the effort required for any specific computational task on this computer to be determined. The large factor between the theoretical minimum value (Eq. (2.10)) and this actual value is due to the fact that in every elementary electronic operation within the processor chip, not just a single electron is transported at minimal voltage, but rather, despite all miniaturization, considerable charge packets must still be shifted. Additional electrical energy is consumed for fans and other assisting electrical computer components.

An unfortunate consequence of this discrepancy is the high demand for electrical power, which is converted into heat worldwide by computers and other electronic systems. This power consumption is on the order of one terawatt, equivalent to several hundred power plants. Therefore, the development of new processor principles with low power consumption is urgently needed, not only for technical and economic reasons but also especially from an ecological standpoint.

Since the energy used is ultimately converted into heat, in the case of computers, as well as all other electrical devices, the increase in "systemic disorder" is directly measurable as entropy. All energy-saving measures and all improvements in the efficiency of electrical devices aim to reduce this entropy production. From the perspective of direct entropy production, the substitution of incandescent lamps—where over 95% of the electrical energy is immediately converted into heat and less than 5% can be used as light—with semiconductor light-emitting diodes, which have a fraction of the power consumption for the same light output, represents significant progress.

However, this evaluation does not consider that the production and use of these optoelectronic light sources require effort in other areas, such as the extraction of raw materials, their processing, and the comparatively very demanding and energy-intensive semiconductor manufacturing process, as well as disposal after the end of their useful life. Life cycle assessments make it clear that these factors must be considered in addition to electricity consumption. Typically, energy consumption during production and disposal, as well as the amount of greenhouse gases released per unit of light output, have to be included in such assessments.

Additionally, it should be considered that other difficult-to-calculate, indirectly effective factors, such as the construction of buildings and transport facilities, packaging, and distribution, must also be reconsidered. The required effort is not only reflected in energy consumption and molecular or atomically dispersed by-products, which can be well quantified with the classical concept of entropy. It also involves the use and distribution of materials for which the arrow of time is equally effective, but for other reference quantities than the Boltzmann constant must be considered.

2.3 Far-Equilibrium Processes and Dissipative Structures

Structures can be found on practically all scales. Many can be directly perceived with our senses, while others can be detected using instruments. Despite the immense diversity in which structures present themselves to us in nature and technology, two large groups can be fundamentally distinguished with respect to the arrow of time (Fig. 2.12a). The structure of a crystal lattice, the houses of a city, and the annual rings of a tree belong to one category, while the flames of a fire and the waves on the sea belong to the other.

What fundamentally distinguishes the annual rings from the ocean waves? The annual rings reflect the periodicity of the annual cycle. Their formation is the result of a periodic process, manifested in the alternation of denser and less dense material embedded in the wood. This structure is created as a result of a physiological process in which order arises by the tree absorbing light energy, emitting thermal radiation, and thus exporting entropy to its surroundings. The resulting structure of the annual rings is conserved. Although the energy-dissipating process is necessary for its formation, the annual rings remain even after the growth and entropy export process has ended. For example, the annual rings can still be visible centuries later in the form of veneer on an old piece of furniture. Through secondary, preserving processes such as mineralization, the tree rings can be preserved as structures in the form of tree trunk fossils for even hundreds of millions of years.

A completely different situation exists with water waves. When the wind blows strongly, large waves build up, but when it dies down, the wave heights soon decrease, and in total calm, the sea becomes mirror-smooth. Like the tree rings, the water waves are an accompaniment to an energy-dissipating process. However, unlike the tree rings, they disappear as soon as their driving force ceases. They are not conserved. Their existence is tied to a continuous driving force, the continuous dissipation of energy. Structures that have this direct dependency on an energy-dissipating process are therefore called "dissipative structures." They are dissipative structures in the narrow sense.[6]

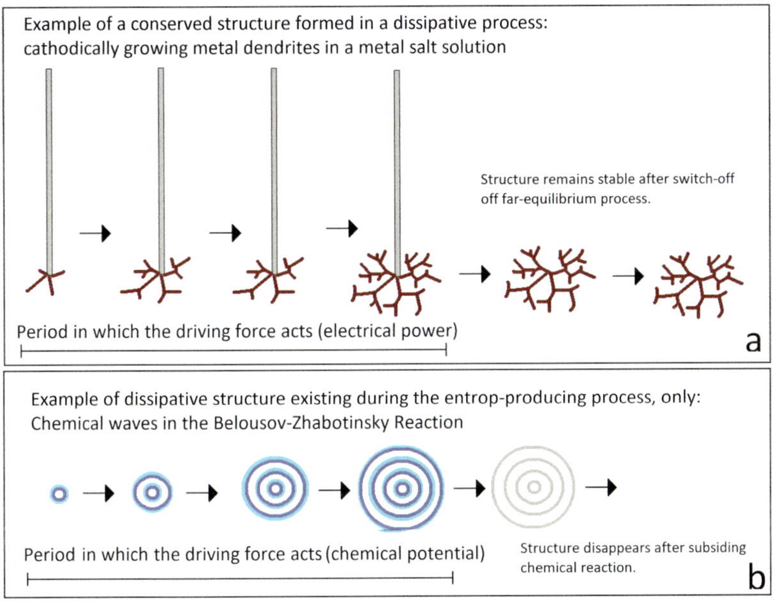

Figure 2.12 Spontaneous structure formation in dissipative processes: (a) electrochemical formation of a dendritic metal structure during the cathodic deposition of metal from a salt solution under high current densities (example of a conserved (frozen) dissipative structure), and (b) formation of propagating chemical waves by the Belousov-Zhabotinsky-Reaction (BZR) in a homogeneous reaction solution (example of a temporary formed (non-frozen) dissipative structure).

Dissipative structures represent a form of transient order. They arise when a system is sufficiently far from thermodynamic equilibrium. As equilibrium is approached, dissipative structures collapse. From a thermodynamic perspective, the continuous flow of energy, the production of entropy, and the constant export of entropy are essential prerequisites for the existence of a dissipative system.

[6]G. Nicolis and I. Prigogine, 1977.

In addition to waves and flames, water vortices and fireworks, living systems are also understood as dissipative structures.[7] Although a living body in its physical composition is at least partially a conserved system, the life processes themselves are a typical dissipative process, which involves the formation of dissipative structures. Concentration distributions of gases and nutrients in cells, tissues, and organs, on one side, and nerve impulses and sound signals on the other side, are typical dissipative structures.

Dissipative structures often manifest as characteristic, often periodic, patterns. An example of this is spontaneously oscillating chemical reactions. In the Belousov-Zhabotinsky reaction (BZR), a redox reaction of malonic acid with bromate catalyzed by cerium ions or ferroin, phases with periodically changing dominance of the reduced and oxidized forms of the catalyst.[8] In a stirred vessel, this periodic change can be observed through color shifts. A complex reaction mechanism with autocatalytic sub-reactions leading to rapid transitions between phases is the cause of such chemical oscillations. When solutions are at rest in the BZR, chemical excitation waves spread out in space. Such concentric, spiral, or helical structures of chemical waves, as seen in the BZR, are examples of chemical dissipative structures (Fig. 2.12b). Instead of propagating waves, stationary dissipative structures, known as Turing patterns, can also form under certain conditions. Similarly, patterns in the arrangement and movement of organisms in swarms represent such dissipative structures, such as in schools of fish or flocks of birds, as well as in the arrangement of mobile microorganisms.[9]

Every structure, including dissipative ones, contains structural information. In this sense, a dissipative structure represents an ordered system. Although entropy is continuously produced by the energy-dissipating process, the dissipative system represents a state of reduced entropy. The key to thermodynamic understanding lies in the export of entropy.[10] The state of order can only arise

[7]A. Turing, 1952.

[8]A.M. Zhabotinsky, 1964.

[9]A. Nakamasu et al., 2009.

[10]W. Ebeling and R. Feistel, 1982.

if the resulting entropy is carried away externally. Dissipative systems must be thermodynamically open and must ensure the export of entropy (Fig. 2.13).

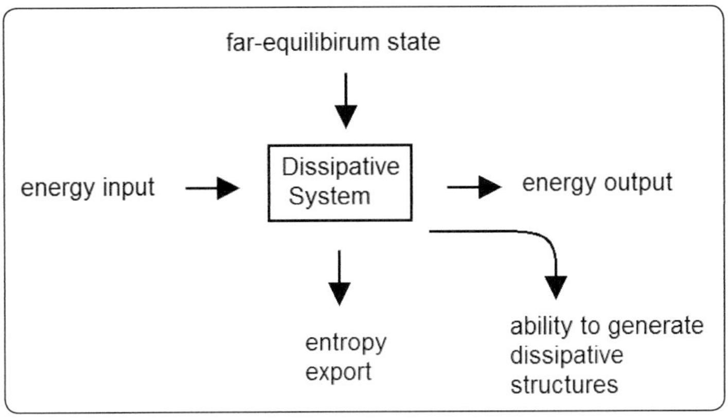

Figure 2.13 The formation and maintenance of dissipative structures require an export of entropy.

The biological formation of shape, the morphogenesis of living beings, is also always tied to the dissipation of energy. Living systems are dissipative systems. Many components of structural formation in organisms can be understood as dissipative structures. The maintenance of life functions requires metabolism and, therefore, the dissipation of energy. For systemic order to be maintained and new orders to be built, living beings must inevitably produce entropy and release it into their environment (Fig. 2.14). Life is always tied to the export of entropy. Interrupting the supply of food or gases leads, sooner or later, in every living being, to an interruption of energy dissipation and thus to death.

Technical processes and systems are in an analogous situation. Technical structures can only be built if energy is dissipated. Technical systems can only operate if entropy is produced and exported. Running technical systems, therefore, just like organisms, are dissipative systems. However, in general, their structures do not collapse when the energy source is turned off. The functional structures of technical systems typically have the character of conserved structures. Users of technical systems expect that they can turn these systems off at will and restart them later. This is only possible if the system remains intact with all its essential functional components, even in the off state. In this sense, technical systems are conserved structures with a switchable entropy-export behavior.

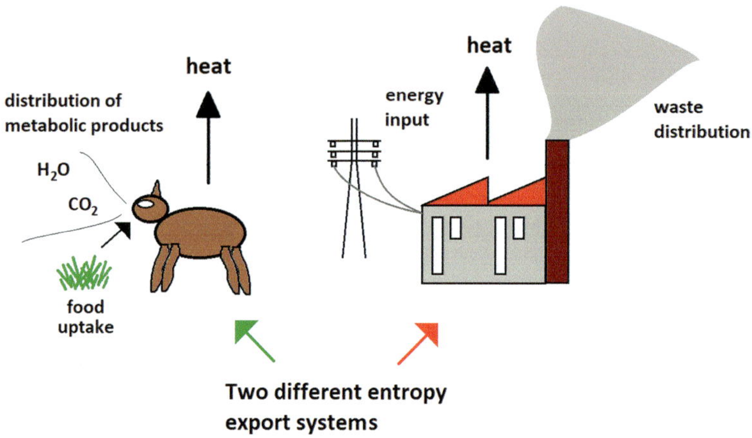

Figure 2.14 Living beings as well as working industrial facilities are systems producing and exporting entropy.

In contrast, higher living beings cannot be turned on and off. At best, the intensity of their entropy export can be reduced. However, there are life forms, such as microorganisms, spores, or seeds, that can exist in a state akin to being switched off, for example, under conditions of low temperatures or dryness.

2.4 Autopoiesis and Gaia Hypothesis from a Thermodynamic Point of View

Life can be accompanied by very intense metabolic processes, but it can also exist under conditions where physiological processes are significantly slowed down. An example of reduced activity is hibernation in mammals. A more pronounced reduction in activity occurs in cold-blooded animals at low temperatures. Some microorganisms can have extremely long cell cycle durations.

Moreover, there are clearly forms, especially in simple single-celled organisms like bacteria, where metabolic processes are completely at rest. This is most likely the case with bacteria that have been trapped in rocks for tens of thousands or even hundreds of millions of years, yet could still be cultivated from their dormant state,[11] as in the case of the salt-loving bacterium *Halococcus dombrowskii*.[12]

Both switched-off machines and completely dormant forms of life, such as spores, have no energy throughput. They do not export entropy. Their structures are not active dissipative but are completely frozen, i.e., conserved.

But what is the essential difference between switched-off technical systems and dormant forms of life like seeds or spores? In the case of mechanical machines, their structure remains essentially unchanged between being turned on and off. The resting state and the entropy-exporting state are similar in terms of the system's structure. The transition between these states does not alter the system, except for the energy throughput, the work the system performs, and the necessary movement to do so. In contrast, the transition between the resting state and the active state in living systems is accompanied by a change in the system's structure and fundamental properties. Typical examples of this are metabolism and morphogenesis. From the dormant form, such as spores or seeds, and with the help of the information stored within them, a new, often very complex, functioning system develops. This ability to "self-organize" is known as autopoiesis.[13] Autopoietic systems thus combine the material and informational prerequisites for transitioning from a conserved, dormant state, meaning a state without any entropy export, into a form- and function-building dynamic, i.e., dissipative, state, provided that external conditions allow for this transition.

---

[11]Dormancy of bacteria; see: A.S. Kaprelyants et al., 1993.

[12]H. Stan-Lotter, 2002; The discovery and re-cultivation of this bacterium by Helga Stan-Lotter supplied a very strong argument for the Panspermie Hypothesis of Svante Arrhenius because it shows the possibility that microorganisms encapsulated in rocks could travel through universe for many millions of years and, therefore, over interstellar and even intergalactic distances.

[13]F.J. Varela et al., 1974.

The principle of autopoiesis from a resting state can be found in bacteria, as well as in the spores of fungi and the seeds of plants. Therefore, it is a principle that is widespread in the biosphere and is by no means limited to primitive life forms like prokaryotes. Ultimately, the development of any organism, any morphogenesis, can be understood as a form of autopoiesis.[14] The development of a complex macroscopic organism from a microscopic form is a typical example of autopoiesis.

Autopoiesis is not limited to individual organisms. The development of forms and functions from simple initial or resting states can also be seen in the development of populations and communities of organisms. A typical example of this is the formation of complex communities during the recolonization of areas after catastrophic events or drastic environmental changes. In these processes, autopoiesis is expressed in the progression from simple pioneer communities to complex climax stages of ecosystems (Fig. 2.15).

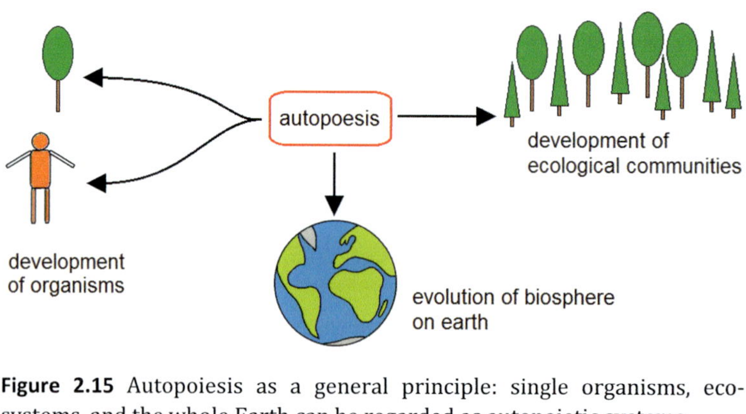

Figure 2.15 Autopoiesis as a general principle: single organisms, ecosystems, and the whole Earth can be regarded as autopoietic systems.

Life on Earth as a whole, i.e., the evolution of the biosphere, can also be seen as a long autopoietic process.[15] The arrow of time, the irreversibility inherent in the evolution of life on Earth, is accompanied by a gradual shaping of global communities. After global catastrophes, which have occurred multiple times

[14]H.R. Maturana and F.J. Varela, 1980.

[15]L. Margulis, 1999.

at the transitions between geological formations, this autopoietic process continued under new, altered conditions, leading to entirely new forms of life and communities. These global autopoietic processes have not only produced the world of living beings but have also altered the global environment, adapting it in many ways to the needs of life. The state of the Earth's upper crust and oceans, the composition of the atmosphere, and the surface temperature of the Earth are significantly the product of these long-term autopoietic processes. The analogy between these processes and the autopoietic morphogenesis of individual organisms has led to the idea of viewing the Earth as a global superorganism. This concept is known as the "Gaia hypothesis."[16]

Like individual morphogenesis, every autopoietic process requires an external drive. Autopoietic systems must be able to take in, transform, and release energy and materials. Autopoiesis requires energy and material throughput. Autopoietic systems are driven; they are dissipative systems. Autopoiesis only works in a state far from thermodynamic equilibrium. Autopoiesis is inevitably associated with entropy production. Autopoietic systems release this entropy to the outside; they use entropy export mechanisms. Autopoietic systems are thermodynamically open systems. Since autopoiesis involves the construction of structures and the generation of information by the system, the entropy export must overcompensate for the entropy production. Thus, autopoiesis always requires an interplay of internal conditions—the system's autopoietic capability—and external conditions—the possibility for energy, entropy, and material exchange. This applies to every organism, every single cell, and even to the Earth and its ecosystem.

2.5 The Earth as Driven System

The light coming from the Sun plays a crucial role for the Earth, for the dynamics on its surface, and especially for life. The interactions of light with matter on the Earth, particularly with living organisms, are highly diverse. One particularly important

[16]J. Lovelock and L. Sagan, 1974.

The energy of a photon is far greater than the energies of thermal excitation at room temperature. This fact distinguishes photochemical processes from other chemical reactions. Without light, only a very small fraction of molecules can undergo a chemical reaction. This fraction is determined by the distribution spectrum of energies during thermal excitation and the height of the activation barrier that must be overcome. The quantitative description of this phenomenon is made possible by the Arrhenius equation for the rate constants of chemical reactions (see Eq. (1.12)).

Photochemical reactions fundamentally differ from these thermal reactions in that, by absorbing the energy of a photon, a molecule receives all the energy necessary to activate the reaction, and in principle, all molecules of an ensemble are capable of this, provided they absorb the photon energy. The absorption of a sufficiently energetic photon by a molecule directly triggers the chemical reaction. This also applies to reactions where the activation barrier is so high that the fraction of molecules that could react thermally becomes vanishingly small. Light thus initiates processes that could not occur otherwise under ambient conditions.

The energy of photons can be well described by their wave nature. It is determined by the wavelength or frequency of the light, with Planck's constant h acting as the coefficient:

$$E_{\mathrm{Ph}} = h \times \nu = h \times c/\lambda. \tag{2.13}$$

A photon of blue-green light with a wavelength of 0.5 μm, therefore, has an energy of approximately 4×10^{-19} J. This is about two orders of magnitude greater than the thermal energy ($k_B T$) at room temperature. In a room-temperature environment, with which the particle is coupled through mechanical collisions, such activation represents an extreme displacement.

In contrast to the mechanical excitation of molecules due to thermally induced mechanical collisions, the state after the absorption of a high-energy photon represents an electronic excitation. Since the energetic state of electrons is crucial for the bonding state within a molecule, the coupling of light energy into the electronic system means that the bonding state is directly altered. This is the basis for the high efficiency of photochemical reactions. In the most favorable case, each absorbed photon directly leads to a chemical reaction. In this case, we speak of a quantum yield (photon yield) of 100%.

The excited state must be considered extremely improbable compared to the non-electronically activated environment. This special status can be expressed as the probability P_{exc} when the energetic expenditure, i.e., the activation by the photon E_{exc}, is related to the thermal energy (reference expenditure):

$$E_{exc} / E_{th} = \ln(1/P_{exc}) \quad \text{or} \quad P_{exc} = e^{-E(exc)/E(th)}. \tag{2.14}$$

This special status can also be explained as the potential for the export of entropy through thermal dissipation. The entropy export reserve $S_{exc-res}$ introduced by the photon can be described by the difference between the two probabilities:

$$S_{exc\text{-}res} = S_{out} - S_{in} = \ln(1/P_{exc}). \tag{2.15}$$

The particle excited by the photon, assuming the excitation occurs at least in the vicinity of room temperature, exists in an extremely non-equilibrium state. From this fact arises the possibility of forming energy-rich compounds, as in biological photosynthesis and many other photochemical reactions. Alternatively, the excitation energy can also lead to the chemical destruction of the dye, known as photobleaching. Or it can simply be released as thermal energy, which is the typical pathway for all dyes that are not involved in photochemical reactions, and is referred to as "non-radiative deactivation." When non-radiative deactivation competes with a photochemical reaction, it reduces the quantum yield of the latter (Fig. 2.18).

The ability of living nature to utilize the entropy export potential gained from light is the central thermodynamic component of the development of the biosphere and, consequently, of the biologically determined Earth's atmosphere through photosynthesis. This ability is the reason for the current, globally speaking, very life-friendly state of our planet.

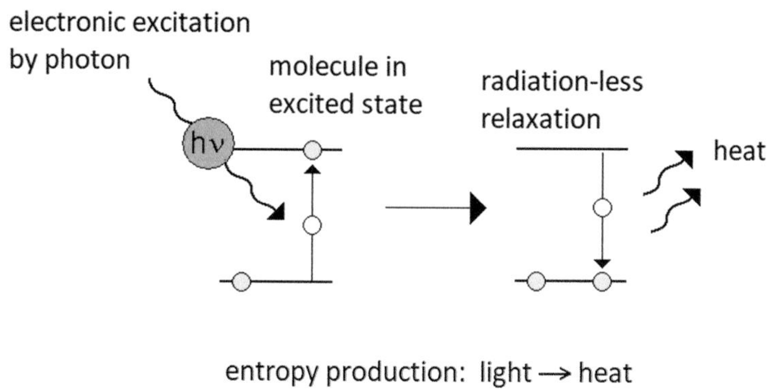

Figure 2.18 Principle of a "radiation-less" (thermal) deactivation of an electronically excited state (thermal relaxation).

Sunlight is largely absorbed by the Earth's surface and the ocean, which is fundamentally independent of whether living organisms are present or not. However, the coupling of the metabolic processes of living nature to sunlight makes the thermodynamic potential of this radiation usable for Earth ecosystems. The entropy balance globally follows the same principles as those considered for a single cell or molecule.

First, let us examine the Earth's energy balance. The Earth is a thermodynamically open system, as it exchanges energy with outer space. Incoming radiation from the Sun is only one side of the thermodynamic system. It is marked by the mean density of radiation absorbed by the Earth, the so-called solar constant (Fig. 2.19). If the Earth were to receive radiation solely from the Sun without releasing any energy into space, it would have long since heated up to the Sun's surface temperature. There would be no liquid water and no solid crust. Instead, the Earth would be a fiery ball of hot plasma.

However, in addition to the visible and invisible radiation that comes from the Sun and is absorbed by the Earth, there is a second radiation process. This process involves the entire Earth's surface and the entire sky. It is not visible to our eyes because it lies in the infrared range of the electromagnetic spectrum. This process facilitates the transport of excess heat into space. The functioning of this process is because, aside from the very minor contribution

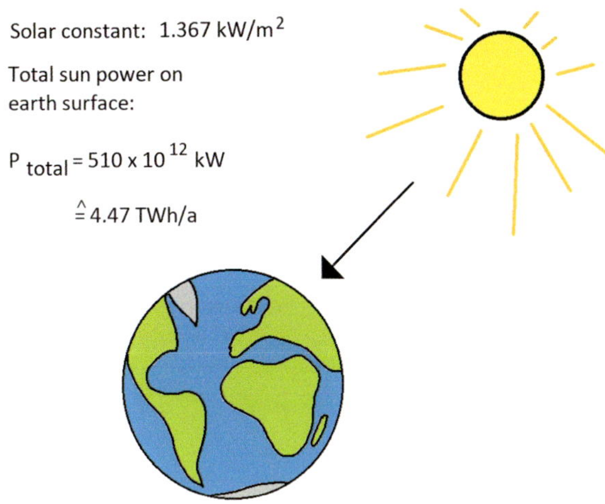

Figure 2.19 The Earth takes up sun energy permanently with a mean energy absorption density of about 1.4 kW/m^2.

from distant gas clouds and stars, the cosmic background emits practically no shortwave radiation to the Earth, but only emits cosmic background radiation, a spectrally broad microwave radiation with a maximum wavelength of about 1 cm. This radiation corresponds to an equilibrium temperature of 2.7 K, which is below the condensation point of helium. Therefore, the cosmic background is extremely cold, and astronauts perceive it as pitch black outside the Earth's atmosphere. The Earth radiates energy into this background. The energy of the photons in this Earth radiation corresponds to the Earth's surface temperature of approximately 300 K, and the wavelength of maximum of this radiation lies in the mid-infrared range (Fig. 2.20).

It is not by chance that the Earth, as a global system, maintains a roughly constant surface temperature over decades and centuries, despite continuous solar radiation. On average, our Earth emits as much energy into the cosmos via infrared radiation as it receives from the absorption of incoming solar radiation. Since the photons from the Sun are much more energetic than the photons of infrared radiation emitted by the Earth, the number of infrared photons emitted per unit of time is much higher than the number of absorbed photons from solar radiation.

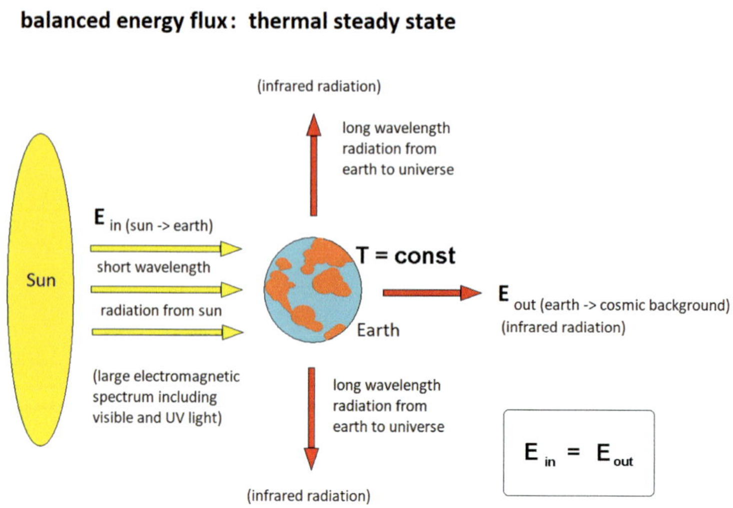

Figure 2.20 Energy balance of a planet with constant surface temperature (thermal steady state).

In fact, the Earth's surface temperature is the result of a self-regulating process. This process works because the radiation power of a surface depends strongly on its temperature. It increases according to the Stefan-Boltzmann law, with the fourth power of the absolute temperature:

$$P_{\mathrm{em}} \approx T_4. \tag{2.16}$$

The intensity of the radiation emitted into space, i.e., the total power emitted by the Earth, is therefore higher the higher the surface temperature is. When the Sun radiates more intensely, the Earth responds to the increased energy input with a temperature rise. This leads to an increase in the power of infrared radiation emitted by the Earth. The surface temperature of the Earth rises until the emitted power once again equals the absorbed power. The system thus stabilizes at an elevated level of temperature and energy throughput. If the energy input decreases due to a less radiant sun, the regulation occurs in the opposite direction. The initially higher surface temperature of the Earth leads to cooling through radiation that exceeds the energy input. This cooling reduces the Earth's radiation power until the emitted power once again matches the incoming power, and a balanced energy budget is reestablished at a lower temperature. A specific average surface temperature of the Earth thus indicates a particular energy balance.

While the energy balance of the Earth, simplified and viewed globally, remains balanced with a constant average surface temperature, the Earth is a significant producer of entropy (Fig. 2.21). If the Earth were a closed system, this entropy production would lead to a continuous loss of order and structure. However, this is not the case. In reality, the Earth is a gigantic entropy-exporting machine.

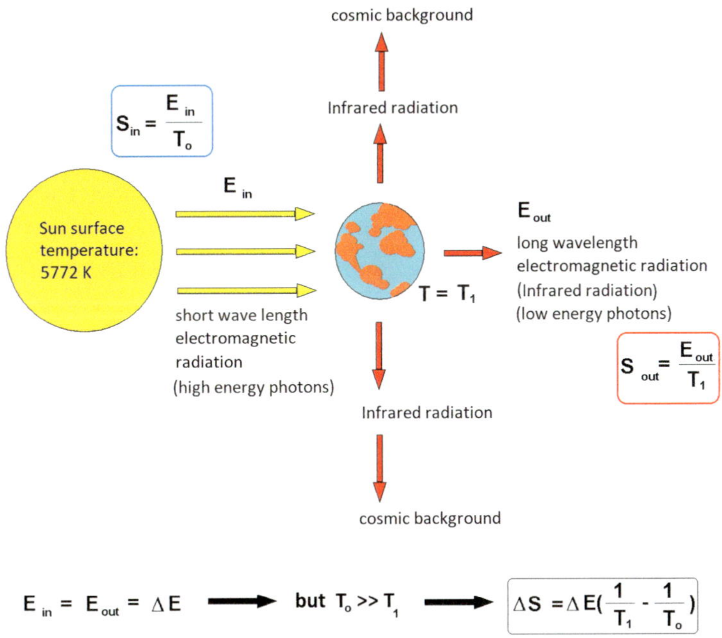

Figure 2.21 The Earth as a giant entropy-producing system and entropy-exporting machine.

The global production and export of entropy are determined by the energy conversion on the Earth. The key factor is the transformation of the shorter-wavelength radiation absorbed by the Earth, which comes from the sun, into the relatively long-wavelength radiation that the Earth emits to the cosmic background.

For a defined period, such as one year, these quantities can be directly expressed as absolute energies and entropies.

For $E_{in} = E_{outc} = \Delta E_{erde}$ (balanced energy budget), the Earth is in a steady state. The entropy export potential provided by the global radiation budget is given by:

$$\Delta S = \Delta E_{earth} \times (1/T_{earth} - 1/T_{sunsurface}). \quad (2.17)$$

Currently, the Earth's atmosphere is in a phase of increasing greenhouse effect due to the rising concentration of greenhouse gases. This means that the state of dynamic equilibrium is disrupted. The Earth's average surface temperature must continue to rise until the enhanced emission of infrared radiation (Eq. (2.16)) caused by the increased temperature ensures that, despite the greenhouse effect, the energy radiated on average matches the energy absorbed from the Sun. As the Earth's surface temperature increases, the potential for entropy export diminishes. This potential becomes smaller the closer the Earth's surface temperature approaches the temperature corresponding to the absorbed solar radiation, i.e., approximately the temperature of the Sun's surface (Fig. 2.22). A temperature increase of 5 degrees, given the Earth's average surface temperature of around 300 K, results in a loss of entropy export potential of approximately 1.5%.

The autopoiesis of living systems is, of course, not determined solely by global entropy export potential. In addition to the thermodynamic boundary conditions, chemical and biological mechanisms are crucial for the dynamics of autopoietic systems. Since most chemical and biochemical processes are temperature-dependent, many processes slow down at low temperatures and can be accelerated at higher temperatures. The ecological problems associated with the enhanced greenhouse effect cause stress for many organisms and communities, as these organisms have adapted to specific temperatures or temperature ranges over the course of evolution. Biochemical mechanisms and the stability or reactivity of biomolecules are aligned with these temperature ranges. The stress associated with global temperature increase manifests itself, among other things, in the shifting of climate zones, vegetation zones, and habitats. The shift towards the poles of the Earth during heating periods has far-reaching consequences that extend well beyond a purely thermodynamic perspective. In the Southern Hemisphere, there is no extensive landmass into which terrestrial species could move from lower to higher latitudes. The shift towards the poles also means that communities are much more exposed to seasonal fluctuations, as these fluctuations are much more pronounced in higher latitudes—consider the extreme effects of the transition between polar summer and polar winter—than in lower latitudes.

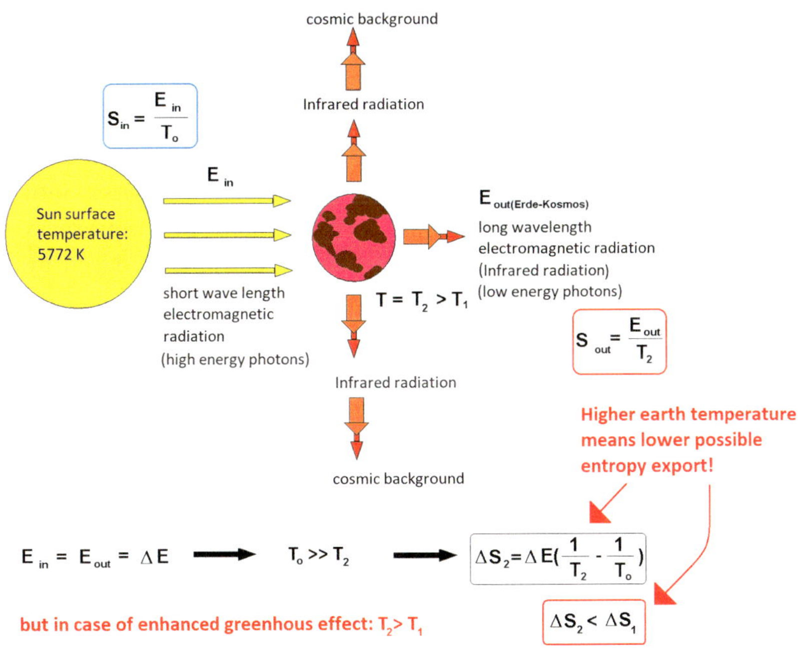

Figure 2.22 Reduced global entropy export potential caused by the greenhouse effect-induced increase of the Earth's surface temperature.

However, stress generally also implies potential. The increasing fluctuations in the Earth's average temperature since the Pliocene (around 6 million years ago) and the more or less periodic shifts between glacial and interglacial phases of the Ice Age have repeatedly led to drastic changes in climate and vegetation zones. On the one hand, these repeated strong changes in habitats have significantly increased the extinction rates of higher organisms.

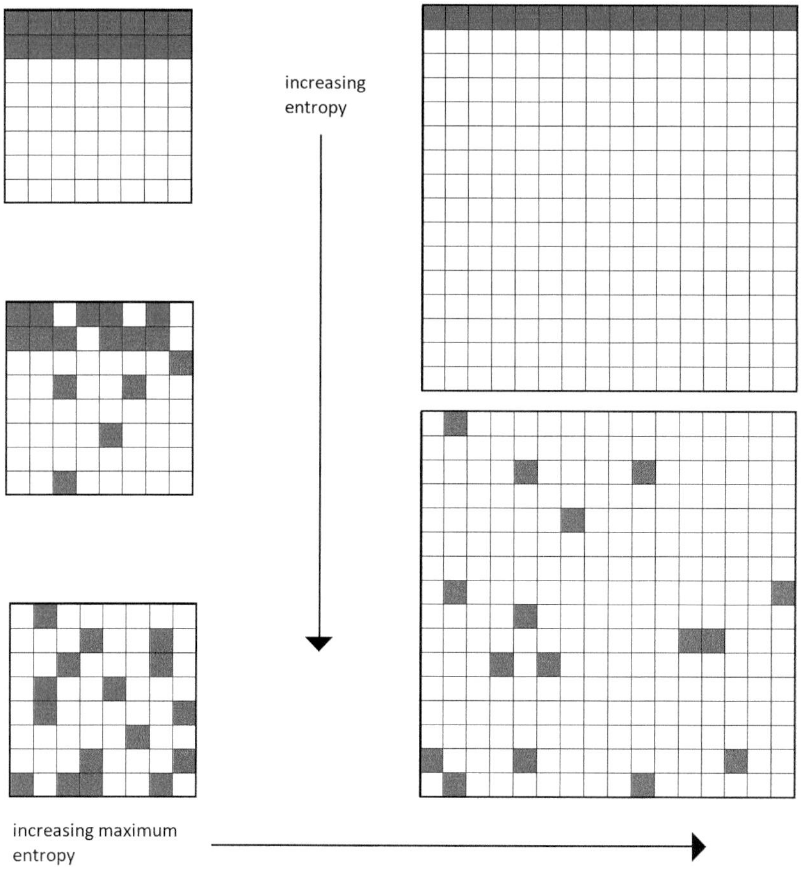

Figure 2.25 Schematic illustration of increasing of maximum possible entropy in the case of an enlargement of a system.

i.e., to improve the selection conditions for the carriers of this genetic material. Such evolutionary advancements concerned small structural motifs, such as patterns of functionally useful amino acids, which led to larger structural molecules through periodic repetition. These multiple repetitions of small structural motifs can be found, for example, in fiber proteins. However, duplications in the genome could also involve entire genes of proteins. Such duplications gave the carriers of genetic information the opportunity to allow variations in the duplicate during further evolution while preserving the original information in the genome. As these processes occurred multiple times during the evolution of genetic material, entire families of genes could emerge within the genome, which are related in sequence but vary in their functions.

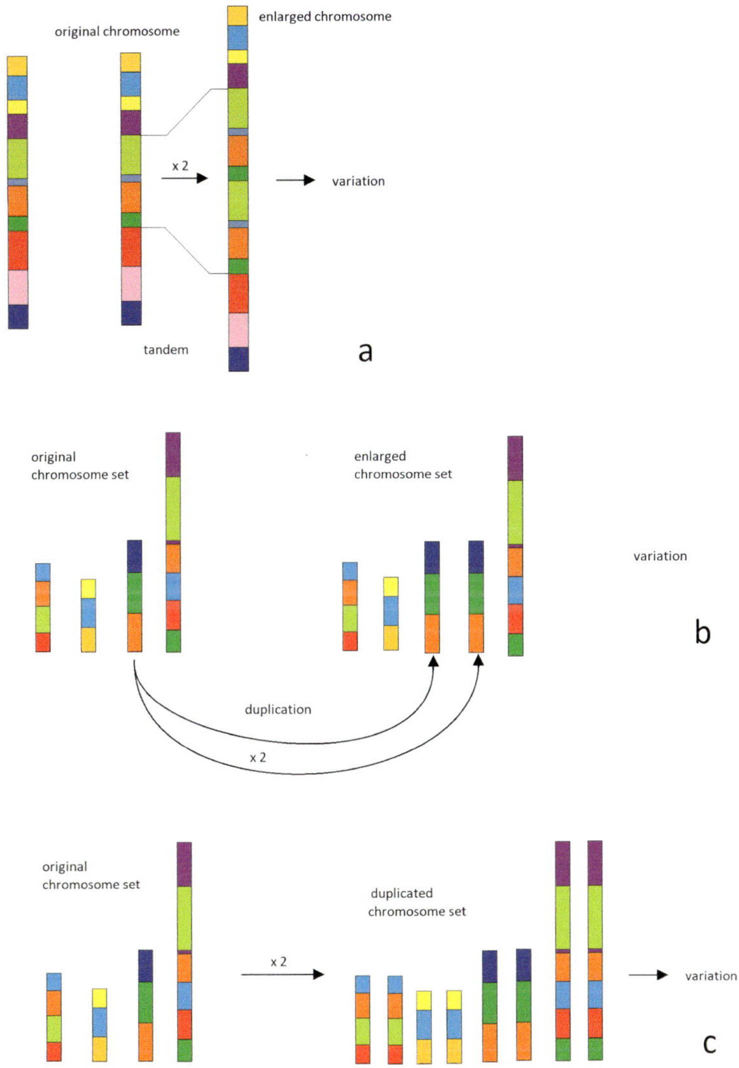

Figure 2.26 Principle of enlargement of genomes by duplications: (a) enlargement of a chromosome by tandem duplication of a sequence section, (b) enlargement of the chromosome set by duplication of one chromosome, and (c) duplication of a complete chromosome set (whole genome).

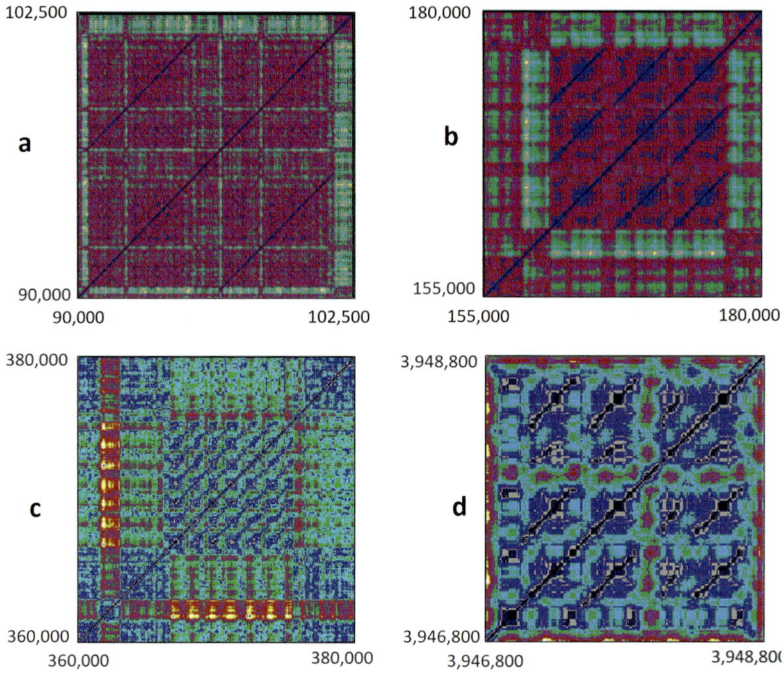

Figure 2.27 Examples of doublet frequency correlation maps of bacterial chromosomes with different numbers of sequence duplications and subsequent mutation densities (the numbers on the axes correspond to the nucleotide positions of bacterial genome): (a) B. subtilis: sequence section of 12,500 nucleotides, (b) B. subtilis: sequence section of 25,000 nucleotide, (c) M. tuberculosis: sequence section of 20,000 nucleotides, and (d) M. tuberculosis: sequence section of 2,000 nucleotides.

Duplications in the DNA sequence are seen as repetition patterns, i.e., translational symmetry. They are also recognizable by a locally altered frequency of elementary motifs, such as doublets. The duplications or periodicities in DNA sequences can be beautifully represented in autocorrelation graphs of doublet frequencies.[20] If the duplications have occurred precisely and have been preserved, this is reflected in a clear repetition pattern as translational symmetry in the correlation graph. A simple duplication is characterized by a range with pairs of parallel diagonal lines (Fig. 2.27a). A little noisy triplication reflected

[20]J.M. Köhler, 2002; J.M. Köhler, 2021.

by three lines is shown in Fig. 2.27b. The correlation graphs show a multitude of parallel diagonal lines in the case of a higher number of repetitions (Fig. 2.27c). However, if small changes, such as point mutations, have occurred during evolution after the duplication, the translational symmetries appear noisy, depending on the number of mutations. The diagonal lines are interrupted or partially shifted in case of stronger noisy sequences (Fig. 2.27d), which means a higher density of small mutations after the tandem duplication.

The price for such an increase in the information content of the genetic material through duplication of sequence segments is, of course, the resulting significant increase in the size of genomes. This could only lead to successful evolution if the efficiency of the molecular machinery involved in DNA replication, essential for cell division, became very high. Long DNA chains had to be copied with great precision. The same applied to the development of the molecular apparatus for the transcription of DNA into RNA and the translation of the genetic code into the amino acid sequences of proteins. However, this is exactly what has occurred in evolution: Even simple bacteria have genomes consisting of several million building blocks. In humans, there are about four billion. And these large amounts of data are transmitted with comparatively high accuracy to daughter cells during each cell division and are continuously translated precisely into proteins in the process of protein biosynthesis.

The entire human genome is distributed across 2 sets of 23 chromosomes. The development of this chromosome set is also involved in several duplication steps in early evolution. Biological evolution clearly shows that the growth of systems, alongside increased entropy, can be beneficial.[21]

A similar process to the evolution of DNA sequences can also be observed in the development of software. Programmers often copy parts of programs and modify subroutines to extend software functions. Such an approach is often much more rational and—cost-effective—than constantly creating entirely new rogram components. The result is—analogous to biological evolution—

[21]P.M. Sharp et al., 1995.

powerful but also very large data packages for our computer programs. Like in biology, programmers also maximize the capacities of ever-evolving hardware to make rapid progress in software development through copying and variation, even though such program growth comes at the price of increased entropy.

2.8 Cosmic Evolution

An analysis of the development of information (negentropy) alongside increasing entropy is also of interest when considering the development of our universe as a whole. Our universe can probably be regarded as a closed system in the sense that there is no exchange of matter, energy, or information with an external environment. The observable universe seems to be a closed-horizon system. However, our universe is not thermodynamically closed, because it is expanding, meaning space is growing. Although the currently most favored model of the universe (the standard model) assumes a constant number of particles (total mass) in the universe, the expansion of space in this model leads to a continuous increase in the maximum total entropy of the universe, $S_{max}(k)$. This results from the growing number of positions in space that can be occupied by individual particles. The following provides an order-of-magnitude estimate based on the number and the theoretically possible number of positions for protons in the universe.

The number of nuclear particles—essentially protons—in the universe, N_{pr}, is estimated to be about 10^{78}. The maximum density, and thus the spatial resolution of the positions, is determined by the extent of the particle by quantum mechanical reason and can be described by the Compton wavelength of the proton, λ_{pr}.

$$\lambda_{pr} = 1.7 \times 10^{-15}\ \text{m}. \tag{2.29}$$

The total number of possible positions for protons in the universe, N_{pos}, can be estimated roughly if we assume a finite extent of the universe, l_k. With the conservative assumption that l_k

corresponds to the extent of the observable universe, about 10^{26} m, we can approximately calculate:

$$N_{pos} \approx (l_k/\lambda_{pr})^3 \approx (10^{26} / 1.7 \times 10^{-15})^3 \approx 10^{123}. \qquad (2.30)$$

Thus, the number of possible positions is many orders of magnitude higher than the number of particles that can be distributed across these positions. The maximum possible entropy in this distribution of protons in the universe can be approximated by:

$$S_{max}(\text{Proton})/k_B \approx \ln\left((10^{123})^{10\wedge 80}\right) \approx 54 \times 10^{80}. \qquad (2.31)$$

Astronomy shows us that the distribution of mass in the universe is far from random, instead, we live in a highly structured universe. The structure hierarchy of mass distribution observed by astronomy confirms this picture: From superbubbles and galactic superclusters to galaxy groups, individual galaxies, star groups, planetary systems, down to moons and comets, structures are found in the universe on scales ranging from hundreds of millions to billions of light years, down to light-seconds (e.g., the distance between the Earth and the Moon). This corresponds to extreme differences in the density of matter, which ranges from one particle per cubic meter in intergalactic space to densities of about 10^{15} g/cm^3 in neutron stars. Homogeneity was recognized only in the intensity distribution of microwave background radiation, in which the relative spatial deviations are in the order of 10^{-5}.

In which direction has entropy developed over the course of the universe's evolution? From what has been said above, it is clear that the maximum possible entropy has significantly increased with the expansion of the universe. But, what does the picture look like for the development of the actual entropy, estimated based on thermodynamic considerations?

According to our current knowledge, an irreversible process is also taking place on a very large spatial and temporal scale. The cosmos itself is undergoing an evolution. If one follows the favored cosmological model, which assumes a constant total mass m_k, the real CMB-related entropy increases slowly due to the decreasing temperature corresponding to the mean wavelength of the microwave background T_{CMB}:

$$S_k(t) = m_k \times c^2/T_{CMB}(t). \quad (2.32)$$

The temperature, in turn, started in the big bang probably with the Planck temperature T_p and the recent CMB-related temperature can roughly be approximated by a square root relation with the time basis expressed by a hypothetical elementary time t_p (Planck time):

$$T_{CMB}(t) \approx T_p/\sqrt{(t/t_p)}. \quad (2.33)$$

This relatively slow increase in CMB-related entropy results from the fact that it is assumed that the entire cosmic mass was already present in a very early phase of cosmic evolution. This is equivalent to assuming high cosmic entropy at a very early stage. This must have been formed in a very rapid and dramatic process. Such an early and rapid development is described by the theory of inflationary cosmic expansion, which is believed to have occurred in a tiny fraction of the first second of cosmic evolution.[22]

According to the standard model, our cosmos is expanding, forming structures, and cooling down simultaneously. If we consider the cosmos as a closed system, then, according to the First Law of Thermodynamics, its energy must remain constant. According to the Second Law, its entropy should increase, since we observe an irreversible process. With constant total energy and decreasing average temperature, cosmic entropy increases. This increase in entropy due to the lower temperature cannot explain why we observe an increasing formation of structures, and thus an increase in information and order—negentropy—in the cosmos. Therefore, the balanced energy budget does not seem plausible on a global scale. The surprising aspect of this observation is that the cosmos is able to cool down and, in

[22]A.H. Guth, 1998.

doing so, form structures. Assuming a constant energy content, the formation of structures—the creation of order through cooling—seems to contradict the Second Law of Thermodynamics. If we do not want to question the validity of the laws of thermodynamics for the cosmos, this means: Either our cosmos is thermodynamically open, or the expansion of space itself compensates for the increasing entropy and even allows an overcompensation of the internal entropy production, which leads to the formation of complex structures and dynamics (Fig. 2.28).

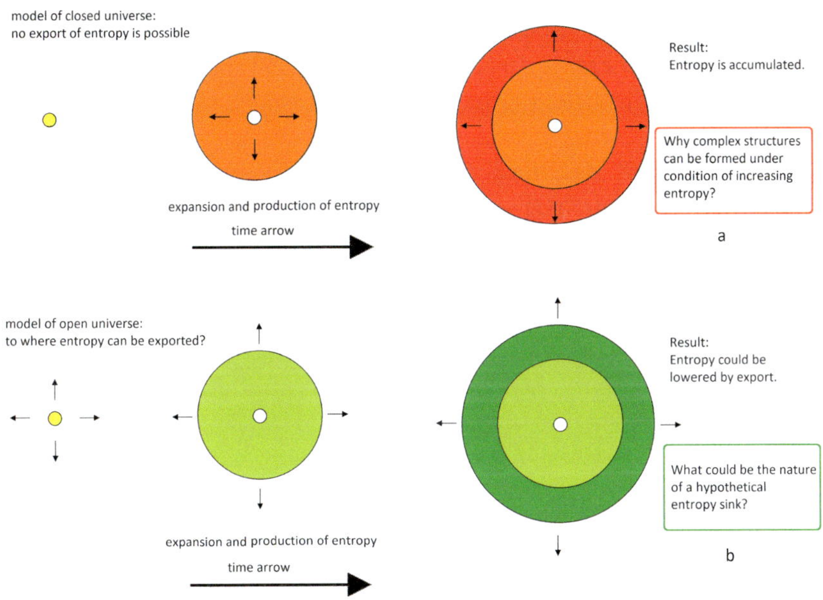

Figure 2.28 Alternative conditions for the development of entropy in the universe: (a) fast accumulation of entropy under the assumption of a closed universe, and (b) reduced increase of cosmic entropy by a hypothetical entropy export path in an assumed thermodynamically open universe.

This point of view corresponds to the idea that the nature of space itself would be responsible for the input of energy into the cosmos and its inhomogeneous distribution, as well as for the consequences that the production of entropy and its inhomogeneous local distribution would have for the cosmos.

A slow entropy in the beginning and a much stronger increase in entropy in the further development of universes would result if one assumes that the mass or energy content of the universe grows from a hypothetical initial mass m_p to its current value over the course of cosmic evolution. Following the cosmological standard model, under the assumption of no adequate mass growth in the cosmos, the expansion of the universe results in a permanent reduction of the matter density in the universe. The fact that comparatively young structures are observed at great distances is an important confirmation of this view.

Unlike conventional chemical systems on the Earth, the average matter density in the cosmos is low, and collision processes are rare. Nevertheless, processes in the cosmos, such as particle interactions during encounters with massive gas clouds, follow statistical laws described by the Boltzmann equation. But, can such a consideration be applied to the cosmos as a whole?

With the simple assumption of linear growth with time increments of size t_p (Planck time), the following equation could be applied for the evolution of cosmic mass:

$$m_k(t) = m_p \times t/t_p \tag{2.34}$$

$$S_k(t) = m_k(t) \times c^2/T_{CMB}(t). \tag{2.35}$$

With such an increase in cosmic mass over time and a simultaneous decrease in temperature, the cosmic entropy would be very small at the beginning. Its initial value would probably correspond to Boltzmann's constant:

$$S_k(t = 0) = m_p/T_p = k_B. \tag{2.36}$$

The temporal development of T_{CMB}-related entropy would then be approximately described by the following equation:

$$S_k(t) \approx m_p \times (t/t_p) \times c^2/T_{CMB}(t) \tag{2.37}$$

or

$$S_k(t) \approx m_p \times (t/t_p) \times c^2 \times \sqrt{(t/t_p)}/T_p = m_p \times c^2 \times (t/t_p)^{3/2}/T_p. \tag{2.38}$$

Thus, cosmic entropy S_k would grow more than linearly.

Today's problems concerning the understanding of the cosmos as a thermodynamically open or closed system are further exacerbated by the assumption of a cosmic evolution from a very small spatial region—the Big Bang hypothesis—due to the fact of an enormously high mass density at the initial moment. As a way out, the model of the so-called inflationary phase was developed, which describes the early phase of cosmic evolution.[23]

Another approach to resolving this contradiction, the steady-state theory, was proposed early on by astrophysicist Fred Hoyle, who considered a continuous input of mass into the universe to be possible.[24] However, all his attempts to prove such a continuous mass input—even today—failed.

The contradictions in the modern cosmic worldview also call for radically different solutions. For example, astrophysicist Hans-Jörg Fahr suggests that an interpretation of redshift that deviates from the expansion of the universe must also be considered.[25]

2.9 Evolution of Systems for the Accumulation and Exchange of Information

Systems with a low capacity for storing information can inherently possess only a limited maximum entropy, as the maximum entropy of a system is described by the maximum possible number of distinct states. Therefore, the Second Law of Thermodynamics is only meaningful for evaluating a state of order as long as the maximum possible number of states remains constant. This applies, for instance, to a space containing a constant number of particles and their principal positions. For such systems of a constant size, entropy export is the only way to reduce internal entropy and increase the state of order.

Living systems are thermodynamically open systems, meaning they also use the option of entropy export. Thus, the fundamental

[23]A.H. Guth, 1998.

[24]F. Hoyle, 1948.

[25]H.-J. Fahr, 2016.

thermodynamic condition for life processes is fulfilled. However, living nature has not only developed the ability to optimize its entropy export but also the ability to increase the number of possible states, i.e., to acquire ever-greater capacities for storing and processing information. As the number of possibilities grows, so does the maximum entropy a system can possess, but at the same time, the space for different systemic states expands as well.

Life and development require the simultaneous existence of random variability and order parameters. Structure, patterns, distinctiveness, individuality, design, and developmental capacity are found neither in a state of minimal nor maximal entropy. However, the room for maneuver between total order and maximum chaos expands with the increasing number of players and the number of positions they can theoretically occupy (Fig. 2.29). The growth of organisms, the enlargement of genomes, the multiplication of body cells during the growth of multicellular organisms, the expansion of populations, and the colonization of new habitats are typical biological mechanisms for expanding the space of possibilities. Biological populations and ecosystems, through growth, expansion, and differentiation, acquire an ever-greater maximum entropy capacity. At the same time, these systems ensure through entropy export that the internal system entropy remains limited, i.e., it occupies only a fraction of the maximum possible entropy. This leads to a scenario where, despite growing entropy, the information content, the state of order, and the evolutionary potential of living systems can also grow.

For simply structured systems, the relative measure of order can also be well quantified. An example of this is the redundancy in the genetic code: with a triplet of four possible nucleotides (e.g., with the DNA bases guanine, cytosine, adenine, and uracil), $4^3 = 64$ different codes can be represented. However, these are used, essentially, to distinguish only 20 amino acids. The maximum entropy of nucleic acid coding $S_{\mathrm{nmax(nuk)}}$ is simple:

$$S_{\mathrm{nmax(nuk)}}/k_{\mathrm{B}} = 3 \times \ln(4) = 5.55. \tag{2.39}$$

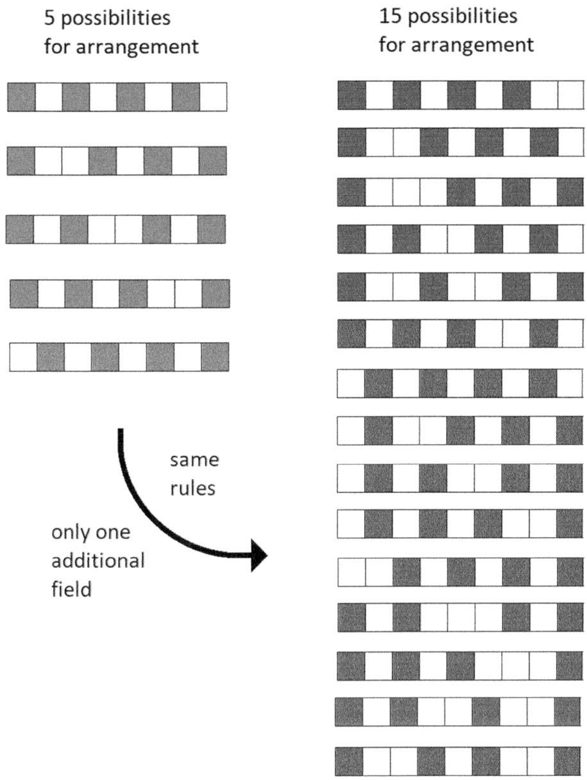

Figure 2.29 Fast increase of variability (maximum entropy) by enlarging systems illustrated by a simple model: Five possibilities of arrangement result if four objects have to be placed apart in a field of eight available positions. The number of possible arrangements grows to 15 in the case of an extension by only one further position to a field of nine possible positions.

The maximum entropy for amino acid selection in protein construction S_{max} (AS) is:

$$S_{max}(AS)/k_B = \ln(20) \approx 3.00. \qquad (2.40)$$

From this, the degree of freedom F_S is derived as:

$$F_S = 1 - S_{AS1}/S_{nuk1} = 0.46. \qquad (2.41)$$

A vanishingly small degree of freedom means that the code allows no flexibility in interpretation. A degree of freedom of 1 indicates that the maximum entropy in the code is transferred 100% into the maximum entropy of interpretation.

In information theory, maximum entropy H_{max} is related to the binary code. For the number of possibilities N, the maximum possible entropy is:

$$H_{max} = \log_2(N). \tag{2.42}$$

Using this value, a relative measure H_{rel} can be defined for sequences, reflecting the deviation from maximum order based on the actual entropy H:

$$H_{rel} = H / H_{max}. \tag{2.43}$$

In a bacterial population of 100 billion cells, the entropy of variability among individual cells is:

$$S_{Zellen}/k_B = \ln(10^{11}) = 25.328. \tag{2.44}$$

The theoretically possible variability in a single bacterial genome, with a genome size of about 4 million base pairs, is:

$$S_{Genom}/k_B = \ln(4^{4,000,000}) = 5,545,177. \tag{2.45}$$

This means that the number of possible nucleotide sequence arrangements in the genome is vastly larger than the assumed population size. Relative to the theoretical variability in the bacterial genome, the entropic degree of freedom in this example is:

$$F_{S(Bakterien)} = 1 - S_{Zellen} / S_{Genom} = 0.999995. \tag{2.46}$$

This indicates that even with the enormous number of cells in the population, only a tiny fraction of the theoretically possible DNA sequences can be realized.

This figure gives an impression of the immense theoretical variability of genomic information even in the simplest living organisms. At the same time, it provides insight into the small fraction of the potential possibilities that can occur in real systems.

This approach can also be applied to other aspects of inheritance, such as those relevant to higher organisms like humans. An estimate can also be made for the role of genetic recombination through sexual reproduction. Let us assume a population of about 100,000 individuals, where—for simplicity—each of the 23 human chromosomes exists in only two variants. The theoretically possible genetic variability is then—greatly simplified—characterized by the number of chromosome variants. The maximum possible entropy of chromosome combinations is:

$$S_{\mathrm{max(chr)}} / k_{\mathrm{B}} = \ln\left((2^2)^{23}\right) = 31.88. \tag{2.47}$$

The actual diversity achievable is determined by the population size:

$$S_{\mathrm{max(pop)}} / k_{\mathrm{B}} = \ln(100{,}000) = 11.513. \tag{2.48}$$

The resulting degree of freedom is:

$$F_{\mathrm{S}} = 1 - S_{\mathrm{pop}} / S_{\mathrm{chr}} = 0.64. \tag{2.49}$$

If we assume a significantly greater genetic variability in the population, with 10 distinct variants for each chromosome, the resulting degree of freedom rises to 0.88, but it remains significantly below 1. For a small population of only 100 members, this figure approaches 0.95, much closer to 1. Due to the high variability within individual chromosomes, the actual diversity in real systems is far higher. Smaller populations are much less capable than larger populations of exploring their combinatorial space of possibilities, which—especially in the face of changing environmental conditions—presents a significant disadvantage in terms of selection.

In biological evolution, systemic information has been accumulated and differentiated. The development of mechanisms for generating, storing, and exchanging information is naturally closely linked to the material carriers of these mechanisms.

Along with the forms of organisms, biochemical functions, and the interaction of organisms in various ecosystems, the mechanisms for storing and exchanging information have also evolved during biological evolution. These systems combine three essential characteristics: first, the construction of structures that provide sufficient possibilities for storage, i.e., allowing for a large potential entropy; second, the actual storage of information and the creation of usable order within numerous possible patterns, i.e., minimizing entropy in this information storage; and third, the ability to connect the information to the evaluation of the environment and to adapt the information through moderate changes. The development of organizational and variation strategies for systemic information that offer optimal degrees of freedom seems to play a central role in biological evolution. These degrees of freedom should not be too close to zero and not too close to one.

The RNA world hypothesis suggests that, in the earliest life-like systems, ribonucleic acid (RNA) molecules simultaneously functioned as action units, i.e., catalysts for biochemical life processes, and as storage forms for system-inherent information (Fig. 2.30). Both functions result from the sequence of four possible building blocks, the nucleotides, within the linearly structured RNA molecules. This sequence represents, on the one hand, a string of symbols that forms the digital molecular information storage, and on the other hand, the pattern in this sequence ensures that the initially linear molecule folds into a very specific three-dimensional geometry, which constitutes the molecule's biochemical functionality. In the phase of the RNA world, there was thus no distinction between information storage, information processing, and chemical function based on different material classes.

According to the RNA world hypothesis, this situation changed due to biochemical interactions between RNA molecules and peptides, which belong to a completely different material class. Part of the functional tasks in early life-like forms was then taken over by proteins formed from peptides, causing the functional role of RNA molecules to shift toward the storage and processing of systemic information. This process can be understood as a spontaneous symmetry break or a spontaneous "dualization process" within early biochemical evolution. The result was early second-generation life-like forms whose functioning was based on the complementarity of molecular information management—primarily through RNA as the information carrier—and molecular action and reaction management—primarily through peptides and proteins.

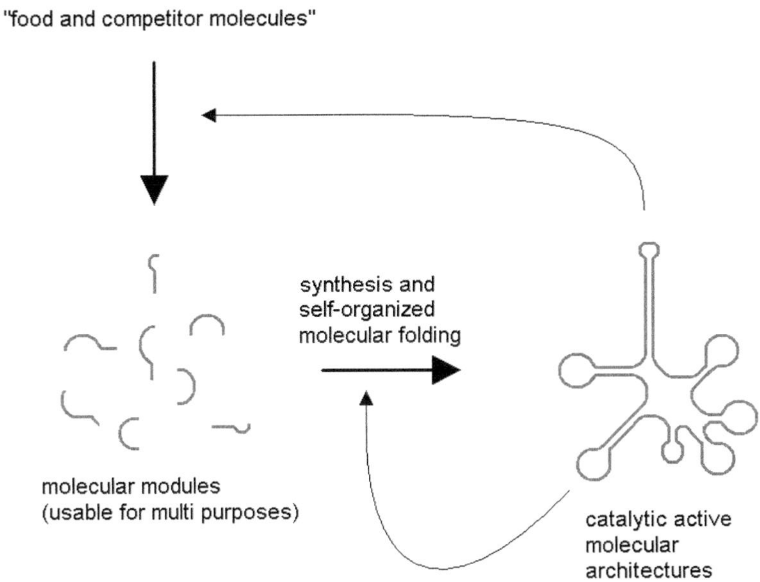

Figure 2.30 RNA world: Small molecular modules (nucleotides) are connected to large RNA molecules, which can fold reproducibly in a stable three-dimensional form, which can catalyze its own synthesis.

If this hypothesis is correct, a second dualization step must have occurred, leading to a division of labor between molecular information storage and the transmission of this information to the action molecules, i.e., the proteins. While RNA retained some biochemical functions, DNA was introduced as a more stable, less hydrolysis-prone storage form for information. This greater stability and, therefore, better suitability as an information storage medium is due solely to the absence of a single OH group

achievements is the development of the nervous system, which is based on electrical excitation states and binary coding. In this system, logical connections are realized similarly to networks of elements (cells), with each element having three possible states (excitable, excited, refractory).

Both archaic, chemically mediated information transmissions and information transfer in nervous systems, along with all other mechanisms, represent principles of order within organisms or communities of life. They depend on metabolic processes, the provision of energy, and the export of entropy. The specificity of information in all cases can be described by the ratio between the total number of possibilities W_0 and the reduced number of possibilities W, as determined by the information.

Ideally, the effort A required for the transmission or processing of information should be minimal. The effort needed for information transfer can be described, by analogy to the equations mentioned earlier (Eq. (2.3)), as the product of a characteristic standard effort A_0 for the respective transmission system and the logarithm of the ratio of possibilities:

$$A = A_0 \times \ln(W_0 / W). \tag{2.50}$$

While for processes on the level of small particles—atoms and molecules—a standard effort is given by the Boltzmann constant, determining the respective standard effort for more complex mechanisms of information transfer and processing is less straightforward.

At the beginning of biological evolution, the principle of identical replication served as the method of information transfer. Systemic information stored in the sequence of basic building blocks within a biological macromolecule was prepared for transmission to the next generation through the duplication of the molecule. This principle is very simple, yet fundamental. It is so fundamental that, once invented, it remained the foundational principle for the transmission of genetic information from a mother cell to two daughter cells throughout the billions of years of evolution and across the entire living world.

The qualitative stages of higher development in evolution can be well identified through the mechanisms of information exchange (Fig. 2.31). They follow a certain chronology of development, from simpler, more fundamental principles to more complex, specialized ones, each building upon the other. For instance, genetic recombination is based on the linear molecular coding of information in nucleic acids and its inheritance during cell division. Nervous systems, in turn, presuppose multicellular organisms with their inheritance systems.

RNA world (hypothetically)	Molecular information storage = molecular tools
	Dualisation: Separation of functions
RNA-protein world (hypothetically)	RNA: information storage/ Proteins: molecular tools
DNA-RNA-protein world (all recent organisms)	DNA: long-term information storage RNA: short-term information storage
Neuronal system	Nucleic acids: slow molecular information processing Neuronal activities: fast electrical information processing
Language	Object-independent information transfer
Writing	Subject-independent informaton transfer
Printing	multi-directional information transfer
World Wide Web	omni-directional information transfer

Figure 2.31 Fundamental steps in the evolution of mechanisms for information exchange.

The development of information exchange systems in human evolution and culture also follows this principle of succession. The fundamental biomolecular and cellular mechanisms, as well as the functioning of the central nervous system of individuals with their diverse sensory and actuation capabilities, remain the foundation of the functioning of organisms. However, new principles of information transfer arose on this basis.

Regarding information transmission, the crucial criterion for the transition from animal to human is the emergence of language. Language enables the exchange of information between individuals using abstract concepts. The objects and processes to which the exchanged information refers do not need to be present during transmission. The place and time to which the information refers do not have to coincide with the place and time of the information transfer. Information is transmitted independently of the subject matter. Place and time, as well as the individuals exchanging information, also become conceptually graspable and thus the subject of abstract description.

The next major leap in the evolution of information exchange systems marks the transition from prehistoric to high culture, accompanied by the invention or introduction of writing. The symbolic materialization of information separates information from the individual. While information exchange through spoken language requires the simultaneous presence of a sender (speaker) and a receiver (listener) of information, writing enables the storage and transmission of information independently of both sender and receiver, that is, independently of the individual subjects involved in the exchange (Fig. 2.32). This allowed information exchange to be separated from the context of a small group of people speaking and listening, thus transcending personal acquaintance, and being practiced among larger, even anonymous groups or societies. Furthermore, information exchange was also liberated in terms of time: the creation of a message and its perception were freed from the constraint of simultaneous sending and receiving. Writing not only enabled subject-independent information transfer but also became the foundation for the cultural storage of abstract information.[26]

The subsequent steps in the cultural evolution of information exchange systems concern the directionality and the technical tools of information exchange. While information written once by an author can be read multiple times and even copied and passed on, it essentially remains a linear transmission of information between individual authors or writers and readers. This situation

[26]G. Höpp, 1970.

changed with the quasi-industrial copying of manuscripts and especially with the introduction of movable-type printing, which allowed the rapid reproduction of written works. With this technology, information dissemination became multidirectional: an author could simultaneously reach a large number of recipients. With the advent of newspapers and, finally, electronic information transmission via radio and television, this principle of multi-directional information distribution became increasingly efficient.

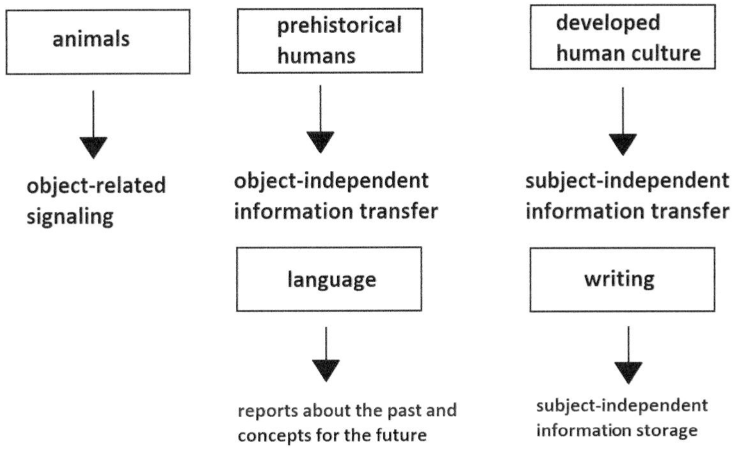

Figure 2.32 Fundamental steps in the evolution of language and the invention of writing concerned the objects and subjects in the information exchange.

Today, the internet brings a new quality to information exchange by allowing not only individual senders (authors) to reach many recipients (listeners, readers, and viewers) but also enabling a vast number of senders to simultaneously inform a vast number of recipients. This global electronic network thus constitutes an "omnidirectional information exchange system."

The information exchanged through writing, printing, and electronic communication also comprises elements of a systemic order. This is especially evident in electronic systems and, more specifically, in the case of volatile storage, where the dissipative nature of the contained structures is highly apparent. All processes of cultural information exchange come with a cost, linked to energy throughput and entropy export. Without the willingness to pay this price—in terms of energy and organizational effort—the mechanisms of information exchange collapse. The throughput of energy and data, as well as the production of heat, are essential prerequisites for modern societal structures. One can certainly assume that the throughput of data, the production of an excess of data (data waste), and the necessity of data filtering are as crucial for the functionality of omnidirectional information exchange as the production and export of entropy are fundamental for any accumulation of information.

It is likely that the filtering effort F, depending on the ratio of the total data volume (i.e., including irrelevant data) d_0 to the relevant data for a specific query d, can be formulated using a standard filtering effort F_0 in analogy to the equations mentioned above (Eq. (2.3), for example):

$$F = F_0 \times \ln(d_0/d). \tag{2.51}$$

It is expected that the further evolution of information exchange systems will initially be shaped primarily by the evolution of filtering mechanisms.

Chapter 3

Entropy, Environment, and Production

3.1 Time Scales of Resource Development, Resource Use, and Natural Recycling

Current efforts to reduce CO_2 emissions and slow down the intensifying greenhouse effect are naturally focused on non-sustainable energy production. The core of the issue lies in the material nature of fossil fuels, which have been predominantly used for energy production since the industrial era. To place their use in an ecologically appropriate context, it is helpful to consider the natural processes that involve, on one hand, the conversion of energy into carbon-rich materials, and on the other, the release of energy through the oxidation of these materials. Both processes occur continuously in nature, independent of human activity, as part of global natural cycles. They operate far from thermodynamic equilibrium and involve the production of entropy. Cycling of all technical products and wastes is an urgent need for achieving sustainability.[1] The process steps in cycling have to be connected with the short-term entropy export processes on the Earth.

[1]K. Webster, 2017.

Entropy and Sustainability: Challenges of Future Energy and Material Management from a Thermodynamic Point of View
Michael Köhler

ISBN 978-981-5129-99-1 (Hardcover), 978-1-003-74889-2 (eBook)
www.jennystanford.com

DOI: 10.1201/9781003748892-4

The most significant natural mechanism for reducing atmospheric CO_2 levels is found in the sunlight-driven photosynthesis of green plants and photosynthetically active microorganisms, primarily unicellular algae and cyanobacteria (Fig. 3.1). These organisms carry out the fixation of carbon from carbon dioxide in the atmosphere or from waters where CO_2 is dissolved as carbonic acid. This natural fixation process reduces entropy within the ecosystem in two ways: First, low-energy starting materials, CO_2 and water, are transformed into high-energy organic compounds capable of releasing energy when oxidized by atmospheric oxygen. Second, carbon distributed in the atmosphere as carbon dioxide is concentrated within organisms. This concentration of carbon also represents an entropy reduction, or in other words, the building of thermodynamic potential.

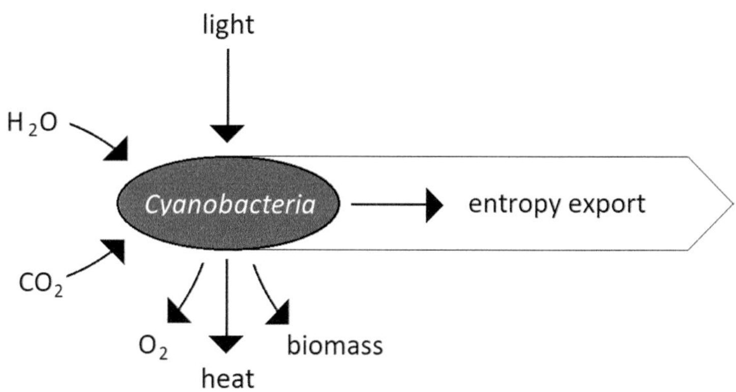

Figure 3.1 Photosynthetically active microorganisms as cyanobacteria, for example, contribute massively to the global biological entropy export system by converting CO_2 into biomass by using sunlight.

These processes are not independent of their surroundings; they only function when coupled with an entropy-export process. The required thermodynamic potential comes from the difference between the energy of photons absorbed from sunlight for photosynthesis and the heat released in the process.

Other organisms, such as humans, depend on the compounds produced by photosynthesis. They meet their entropy export needs by consuming these compounds and oxidizing them with oxygen in respiration. The interplay of photosynthesis and respiration within the global ecosystem maintains a largely balanced material cycle. The carbon dioxide released by respiration into water and air is, on average, taken up by organic substances through photosynthetic activity.

The average residence time of CO_2 molecules involved in these processes in the atmosphere is about 20 years. The mean residence time of carbon in soil and plant cover, however, varies greatly with temperature and other climatic factors. In tropical regions with high turnover, it averages around four years, increases with latitude, and reaches approximately 400 years in the periglacial zones of polar regions.

The use of fossil fuels is, in this sense, a process highly compatible with the global ecosystem, as it seamlessly connects to the natural processes of CO_2 formation and its subsequent reuse in photosynthesis. Ultimately, the use of fossil resources, like that of renewable organic energy sources such as wood, rapeseed, or corn, is integrated into the Earth's global entropy export system, wherein shortwave sunlight is absorbed, and the energy is released as longwave infrared radiation into space. Viewed this way, the use of fossil fuels is actually a very environmentally compatible strategy.

The problem lies in the very long timescales over which carbon accumulates and fixed carbon is deposited in the form of coal, oil, and natural gas, compared to the much shorter period over which humans are utilizing these resources.

A long-term fixation of carbon, meaning binding it over geological timescales, occurs through the deposition of carbon-containing materials in sediments (Fig. 3.2). Natural release from these sediments is tied to the movement of the Earth's crust and thus to tectonic processes associated with plate motion. Only through thermal and chemical stress—primarily in connection with the subduction Earth's crust plates, that is, the downward movement of sediment layers into regions of the Earth's mantle—is carbon from these sediments released again, predominantly as CO_2, which is then emitted into the atmosphere. This explains the lengthy characteristic timescales on the order of several hundred million years.

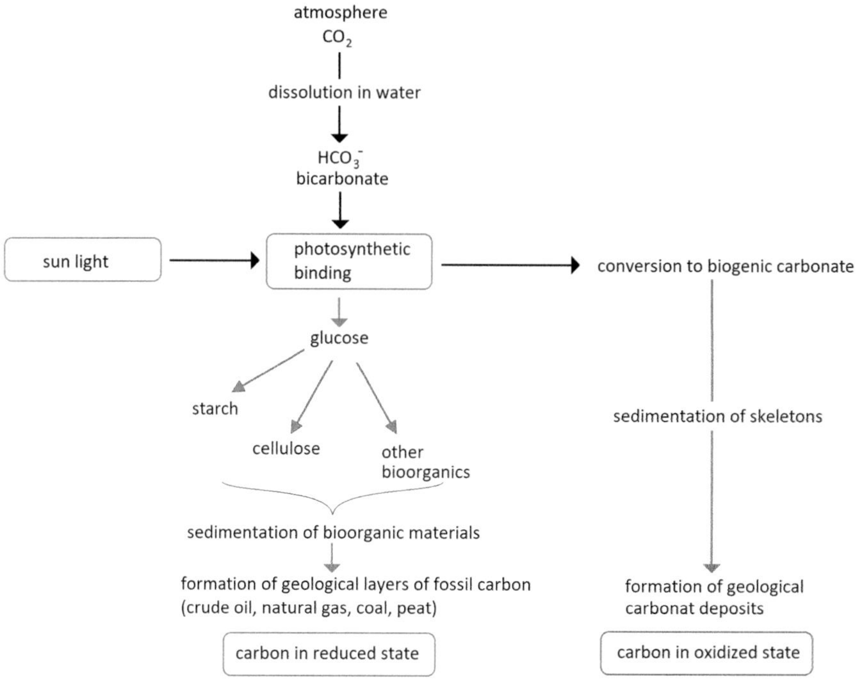

Figure 3.2 Natural reduction of CO_2 content in the atmosphere by sunlight-driven biological and following biogeochemical process chains, resulting in a transfer of atmospheric carbon into the long-term carbon storage in sediment deposits.

Within this long-term carbon cycle, there are two main branches. The first involves the binding and storage of carbon in its oxidized form, i.e., as carbonate. This form of carbon is found in limestone and dolomite, for instance. These materials, which occur in vast quantities within the Earth's crust and constitute entire mountain ranges, are products of biological activity and sedimentation. Stored as carbonate, starting from atmospheric CO_2, these rock formations remove carbon from the Earth's atmosphere and biological metabolic processes.

The second branch involves the binding of carbon in its elemental form, as hydrocarbons or transitional forms, specifically in sediments where carbon is present in a reduced form. This type of carbon fixation also occurs over long geological timescales. Unlike the carbonate form, the organic material produced by photosynthesis is not oxidized but reaches oxygen-poor soils and waters in reduced form, entering sediments. These sediments form the deposits of fossil fuels we use today. Natural release from such sediments, as with carbonate sediments, occurs through tectonic processes related to plate tectonics, which operate on timescales of hundreds of millions of years.[2]

The material energy carriers bound within coal, oil, and natural gas deposits millions of years ago are indirect products of ancient photosynthesis. In previous geological ages, plants and microorganisms absorbed sunlight, using it to convert low-energy carbon dioxide into high-energy organic compounds. This process utilized an entropy-export pathway, resulting from the absorption of shortwave, high-energy photons from sunlight and the release of longwave, lower-energy infrared photons into the cold vastness of space. Consequently, today's fossil fuel deposits represent reserves of entropy export, created long ago and often formed over hundreds of thousands of years.

The temperature difference between the Earth's interior and its surface also constitutes an entropy sink that can be harnessed for entropy export without consuming raw materials or producing environmentally harmful products. This is already an important energy source today, where geothermally heated water or magmatic masses approach the surface. This potential can be exploited in volcanically active regions, such as Iceland, where geothermal power plants make use of it.

3.2 Time Scales in the Evolution of Atmosphere and Climate

On larger time scales, significant changes in the Earth's atmospheric CO_2 levels have occurred. During the Ice Age (Pleistocene), for example, the CO_2 levels fluctuated substantially, ranging from approximately 200 to 350 ppm (0.02–0.035%) in sync with the glacial cycles. These drastic fluctuations coincided with shifts between extensive global glaciation and interglacial warm

[2]Besides green plants, microorganisms play an important role in carbon fixation, for example, in marine and lake sediments (S. Dyksma et al., 2016; A.L. Santoro, 2013).

periods, spanning roughly 100,000 years. Going back in geological history—about 100 million years to the Mesozoic Era—we find prolonged periods with atmospheric CO_2 concentrations exceeding 1000 ppm, over time spans of tens to hundreds of millions of years. The strong variations in CO_2 levels and their connection to these extensive time frames cannot be explained solely by shifts in the balance between photosynthesis and respiration. Rather, these fluctuations are driven by additional biogeochemical cycles involving carbon, whose characteristic time scales are determined by geological rather than biological processes. Thus, a hierarchy of time scales of natural processes must be reconsidered in the discussion of stability and changes in atmospheric CO_2 content and climate change (Fig. 3.3).

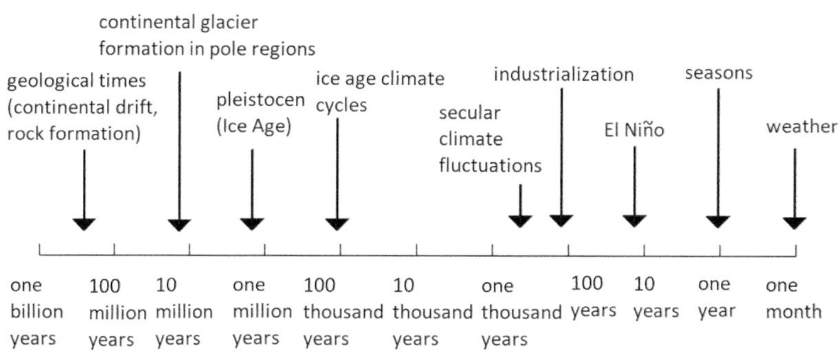

Figure 3.3 Hierarchy of time scales in the development of atmosphere composition and climate on the Earth.

Although the atmosphere is highly dynamic both mechanically and chemically, it responds relatively slowly to some changes in its composition. This lag is due to the differing rates at which substances are broken down in the atmosphere through natural processes or removed through interactions with the oceans and biosphere. Besides their average residence time, the stability of environmental conditions also depends on the rates at which individual substances are produced, as well as where and by what mechanisms they are formed. For instance, the lifespan of ozone in the troposphere is only a few days to weeks. By contrast, stratospheric ozone, which is depleted by ozone-depleting gases like mixed halogenated hydrocarbons in higher latitudes during winter, only partially regenerates through sunlight in summer. The longevity of stratospheric ozone is highly dependent on ozone-depleting radicals, which exist in low concentrations (sub-ppb levels, or less than 1 part per billion for most relevant substances), but have average residence times ranging from years to decades. For instance, the residence time of trifluoromethane is around 270 years,[3] while sulfur hexafluoride—not an ozone killer, but an extremely potent greenhouse gas used in semiconductor technology—lasts about 3200 years.[4] Since ozone-depleting gases and their radicals act as catalysts in the ozone depletion process, they remain unchanged and only trace amounts are sufficient to severely thin the stratospheric ozone.

Most of these gases are not produced naturally and therefore were not present in the atmosphere prior to industrialization. Since the technical removal of these trace gases from the atmosphere is impossible, their release signifies a deviation from the natural state, with effects on the greenhouse effect, climate, and ozone levels persisting over decades to centuries, potentially even millennia, due to their residence times.

The impact of greenhouse gases on climate is primarily determined by their atmospheric residence time and the quantities released. Carbon dioxide (CO_2) is especially relevant for the climate due to its significant contribution. Although its greenhouse potency per molecule is only about a thousandth of the halogenated hydrocarbons mentioned earlier, CO_2's far greater concentration in the atmosphere makes it about a thousand times more significant for the greenhouse effect. With an average atmospheric residence time of about 20 years, CO_2 contributes to a relatively slow atmospheric response to anthropogenic influences. Its climatic impact is further extended through atmospheric interactions with water bodies; the dissolution of CO_2 in water and its transport within the oceans buffer a substantial portion of atmospheric CO_2. However, this process, along with the geological fixation of CO_2 as biogenic or inorganic carbonates in water bodies, is much slower, resulting in a climate impact scale for CO_2 that spans not only centuries but millennia.

---

[3]https://ethz.ch/de/news-und-veranstaltungen/eth-news/news/2015/09/fckw-ersatzstoffe-gefaehrden-das-klima.html.

[4]https://www.umweltbundesamt.de/themen/verbot-des-klimaschaedlichen-sf6-in.

As an upper limit for the time scale shaped by the residence times of greenhouse gases like carbon dioxide and methane, considering long-term feedback with the oceans, the period of glacial climate fluctuations can be used. Atmospheric fluctuations in CO_2 and methane concentrations, identified with high resolution from ice core samples, correlate strongly with long-term temperature fluctuations associated with the transitions between Ice Age stages (glacial) and the interglacial warm periods. These cycles have a period of about 100,000 years. Besides greenhouse gases, other factors also contribute to the periodicity of glacial climate fluctuations, including Earth's orbital movements, the dynamics of the biosphere, and likely geological factors as well. On the scale of glacial climate cycles, the precessional movement of Earth's axis, with a cycle of 26,000 years, about one-fourth of the glacial periodicity and known as the "Platonic Year," plays a significant role.

Conversely, glacial climate fluctuations themselves establish a timescale crucial for the development of the biosphere and atmosphere. The periodic glacial shifts in average annual temperature, sea level, precipitation patterns, and vegetation zones drastically impact biogeochemical cycles. The changing local and global conditions for material transport play an important role in this process. These include large-scale atmospheric currents and ocean circulation, as well as altered conditions for erosion, soil formation, and the associated redistribution of minerals and sediments.

Ecosystems are naturally interconnected with geological processes and climate change. However, they develop their intrinsic dynamics that interact with geological and climatic processes, creating unique timescales. These timescales manifest, for example, in the processes of recolonization of land surfaces following dramatic events like floods, permanent sea regression, wildfires, and volcanic eruptions. Depending on climatic conditions, particularly temperature, precipitation, and soil conditions, recolonization can occur within a few decades or take thousands to tens of thousands of years. In semi-arid regions, soil formation and the establishment of vegetation cover generally progress very slowly, whereas, these processes are much faster in tropical and temperate regions with distinct seasons and sufficient moisture. Even so, in such regions, the development of a habitat from pioneer vegetation to its mature stage (climax) may still take centuries, and in some cases millennia. This extended timescale arises, for instance, from the long generational cycles of certain organism groups and the exceptional longevity of key organisms within ecosystems. In some cases, such as with certain fungi and trees, individual organisms relevant to ecosystem stability can live for thousands of years.

The dynamics of natural ecosystems are typically characterized by the hierarchy of timescales (Fig. 3.3). This hierarchy begins with the short generation times of microorganisms, which can be as brief as under an hour, allowing large populations to develop from single cells within just a few days. Other microorganisms and small organisms, such as insects, have generation times ranging from days to weeks. Larger organisms follow an annual rhythm or, like many vertebrates and woody plants, have generation times spanning years to decades. Since the establishment of populations and complex communities generally requires many generations, it is clear that the typical timescales on which ecosystems respond to environmental changes through adaptation range from centuries to millennia.

Beyond glacial climate fluctuations, there are geological timescales associated with the long-term dynamics of ice sheets. Glaciation in Antarctica began in the mid-Tertiary, over 20 million years ago. While glaciers in high mountains change in size with secular climate fluctuations (over several centuries) and global glaciation levels fluctuate on the timescale of the glacial periods, i.e., about a hundred-thousand-year scale, the formation and dynamics of large continental ice sheets occur over millions of years.

The Ice Age (Pleistocene) is seen by geologists as covering the past 2.7 million years. This period corresponds to cycles of glaciation and ice advances in the higher northern latitudes. However, the overall cooling of Earth's atmosphere has largely been determined by the glaciation of polar continental masses, particularly in Antarctica. The onset of Antarctic glaciation marks the shift from a long-standing warm phase in Earth's geological history to a cooler phase extending into the present. This cooler phase began in the Oligocene, in the mid-Tertiary, and has now lasted significantly longer than the geologically defined Pleistocene. The primary cause of this long-term climatic shift is attributed to the polar position of the large Antarctic landmass. However, the location of this landmass is not fixed when viewed over a sufficiently large timescale; it, too, is subject to continental drift. The timescale of these movements of Earth's crustal plates is on the order of hundreds of millions of years, an order of magnitude beyond the duration of the current Antarctic glaciation. In fact, the Earth has experienced several glaciation phases, such as in the early Proterozoic over two billion years ago, and in the late Precambrian around 800 million years ago. Between these glaciations were long periods characterized by high temperatures and a mostly ice-free Earth's surface.

Not only Earth's geological structures, glaciers, and the distribution of seas and landmasses, but also individual deposits and habitats are the result of spontaneous order formation, for which energy has been applied and entropy exported in various past periods. Every functioning ecosystem, meaning each robustly dynamic system, and every deposit, defined as a space with a specific material separated from other substances, constitutes an ordered structure, which can be understood as a reserve for entropy export. This concept also applies to the global ecosphere, which encompasses landscapes, ecosystems, soils, and deposits, and an atmosphere shaped by its gas composition, biologically influenced, and formed through the Earth's long-term biological activity.

3.3 The Massive Time Scale Violation by Industrial Energy Production Using Fossil Carbon Resources

With the economic use of deposits and surface areas on one hand, and the transformation and destruction of habitats along with alterations in the composition of water bodies, soils, and air on the other, humanity is disrupting these ordered structures and entropy-export reserves that often took geological timescales to form. These interventions are almost always unsustainable, representing irreversible changes. Only the natural processes that initially created these ordered structures could counteract the effects of human interference. This fact, along with the reality that actions should be guided not by the aim of restoring past conditions, but rather by compatibility with a self-regulating global ecosystem, highlights the discrepancy between the timescales of the formation or regeneration of natural entropy-export reserves and their exploitation or destruction as a meaningful criterion for their critical assessment.

For the current economic, social, and political situation, it may seem initially irrelevant whether individual unsustainable behaviors impact timescales of hundreds, thousands, hundreds of thousands, or even hundreds of millions of years. However, the problem of unsustainable practices appears vastly different when evaluated against humanity's long-term survival prospects and the survival of the biosphere. It is anticipated that, with increasing wealth, education, and global responsibility in the world population, attention to long-term development issues will gain political importance. Economic and political systems, as well as individual lifestyles, consumption habits, and personal responsibility, must take this into account. Responsible behavior requires understanding the long-term consequences of using natural resources and shifting the global state. In terms of planning and sustainability, it makes a significant difference whether Earth's ecosystem has the chance to correct human mistakes within centuries, only after millions of years, or not at all if irreversible processes are set in motion that cannot be corrected even over very long periods.

Some climate researchers believe that significant disturbances to Earth's energy balance caused by changes to atmospheric composition could have far more critical and irreversible consequences than "just" a few degrees of global warming, the melting of polar ice caps, a drastic rise in sea levels, and shifts in climate zones. The most dramatic scenario for Earth's global ecosystem would be a continually accelerating greenhouse effect. This could lead to Venus-like conditions for our atmosphere:

boiling oceans, massive amounts of water vapor in a blisteringly hot atmosphere, unimaginable storms, and temperatures far above water's boiling point, making life impossible.

Contrasting such a frightening scenario is the understanding that, throughout the last hundreds of millions of years, the biological components of Earth's ecosystem have repeatedly had a stabilizing effect on the climate. Over the long term, biological activity has steadily reduced the greenhouse effect by decreasing greenhouse gases in the atmosphere. This long-term reduction has been extraordinarily important because, over the timescale of billions of years, the Sun has gradually emitted energy with increasing intensity, raising the solar constant or the energy received per square meter of Earth's surface per year (Fig. 3.4). Without the reduction of the atmospheric greenhouse effect by the biosphere, the Earth might already have become a "running greenhouse," with life-hostile, extreme heat conditions.

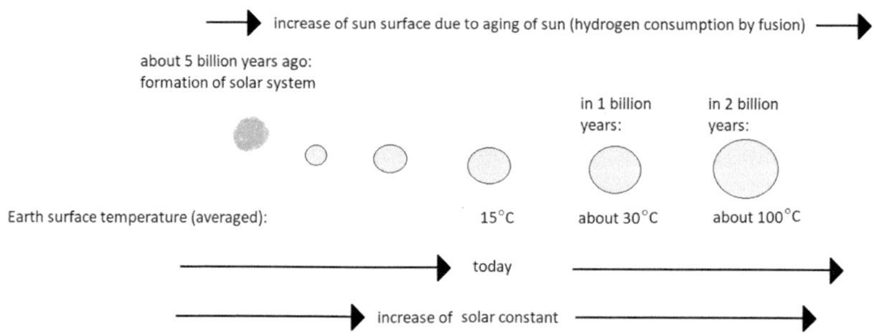

Figure 3.4 Very-long-term heating of the Earth due to aging and gradual expansion of the Sun.

The atmospheric composition developed and maintained by living nature represents an ordered state of fundamental value to Earth's ecosystem. Serving as the foundation for sustaining life-friendly conditions on the Earth, these atmospheric conditions embody a state of reduced entropy. They are the result of a long-term entropy export from the global atmospheric system. This entropy export has been achieved by living nature, as it organized the necessary entropy flow through the reception of solar energy, photosynthesis based on this energy, the global material cycles it enabled, and the thermal radiation emitted back into space.

Particularly noteworthy in this context is that Earth's ecosystem has maintained life-supporting atmospheric conditions by balancing too much and too little greenhouse effect, despite the long-term trend of increasing solar radiation steadily pushing conditions in one direction. On one hand, CO_2 and even more potent trace gases in the atmosphere were reduced sufficiently to prevent overheating. On the other hand, an adequate greenhouse effect was maintained, ensuring that Earth's average annual temperature remains around 15 °C, about 30 degrees higher than it would be without any greenhouse effect, which would otherwise lead to a complete freezing of Earth's surface.

A significant indication of this "fine-tuning" of Earth's average temperature, controlled by greenhouse gas concentrations, is the relatively small amount of CO_2 on which ecosystems, particularly photosynthetically active organisms that rely on CO_2 for their metabolism, survive. It is worth considering that we humans, as oxygen-breathing organisms, have constant access to air with around 21% oxygen content, which we can draw from amply. In contrast, green plants and microorganisms that perform photosynthesis now rely on an atmospheric CO_2 concentration of only around 0.04%, from which they must draw all their carbon to build organic matter (Fig. 3.5). Whereas, oxygen molecules remain in the atmosphere for over 20,000 years on average before being taken up by living organisms, CO_2 molecules must be cycled within a few decades, on a timescale shorter than the lifespan of some organisms and significantly shorter than the developmental timescale of many ecosystems. This "fine-tuning" demands the utmost respect for the complex functionality of the coupling between the biosphere and the atmosphere.

Evidently, an important development in greenhouse effect regulation has taken place over the past two hundred million years. During the long periods of the Mesozoic (the dinosaur era) and the early Tertiary, photosynthesizers experienced atmospheric CO_2 levels exceeding 1000 ppm, providing relatively ideal conditions for extracting carbon from the atmosphere. However, during the glacial phases of the Ice Age, they had to survive with less than 200 ppm. Even in the warm phases between ice ages, the maximum CO_2 concentration in recent geological history has probably not exceeded about 400 ppm, meaning that over the past hundreds of thousands of years, the climate was regulated by all biological metabolic activity at this remarkably low CO_2 concentration.

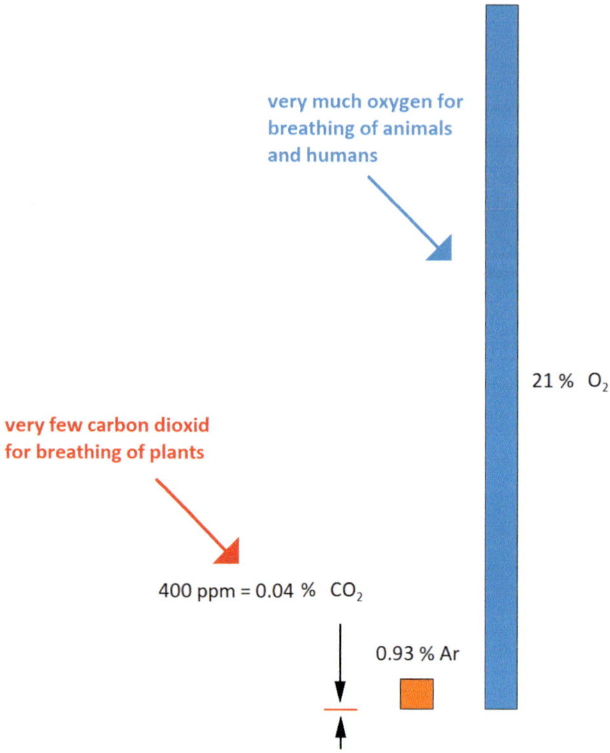

Figure 3.5 Comparison of the content of oxygen, argon, and CO_2 in the recent atmosphere; green plants can use a very small concentration of CO_2 for the generation of organic substances by photosynthesis.

The currently low CO_2 content, viewed on a longer time scale, i.e., with a perspective spanning hundreds of millions of years, ensures that despite high solar output, global temperatures remain relatively tolerable. At the same time, it still allows a sufficient CO_2 supply for the biological photosynthesis systems necessary for organic material. It can be assumed that, over the long term, nature must continually develop more efficient mechanisms for removing CO_2 from the atmosphere so that the atmospheric CO_2 content can gradually be lowered to compensate for the increasing solar output. Additionally, the reduction of ocean surface area and the binding of large water masses within polar glaciers, enhanced by a high albedo effect, meaning a high reflection of solar radiation, is a long-term mechanism that contributes to stabilizing the climate as solar output rises.

The state, which has dominated Earth's climate for approximately two and a half million years, marked by alternating warm and cold periods during the current Ice Age, might represent a still insufficiently stabilized interaction between cosmic and abiotic atmospheric events and the metabolic processes of the biological components of the global ecosystem.

In any case, during the Ice Age, compared to climatically calmer periods in Earth's history, the rates of extinction and evolution among organisms drastically increased, which likely contributed to the development of new ecosystems better adapted to the altered global conditions. It appears that ice ages acted as accelerators of evolution, enabling the replacement of one global ecological regulatory regime with another, involving new types of organisms. The time scales of these changes span millions of years.

Human interventions in the natural environment, whether consciously or unconsciously carried out, and the depletion of the entropy-export reserves accumulated throughout Earth's history, impact a broad range of timescales. The use of energy reserves illustrates this clearly: the utilization and potentially even the overuse of forests, for instance, affects a timescale tied to their growth, on the order of centuries. The use of Quaternary peat, as it was extensively mined in historical times in the North Sea mudflats, involves Quaternary peat formation timescales, spanning a few millennia. The exploitation of Ice Age wood and peat deposits concerns hundreds of thousands of years-old resources. The exploitation of Tertiary lignite deposits concerns deposits formed tens of millions of years ago, while the exploitation of coal typically formed in the Devonian, Carboniferous, or Permian periods spans timescales between formation and exploitation of hundreds of millions of years.

When considering the timescale of resource depletion, one must also consider the timescale of CO_2 release into the atmosphere. Since the beginning of industrialization, atmospheric CO_2 levels have risen from around 270 ppm to over 400 ppm. CO_2 fluctuations between roughly 200 and 350 ppm are typical for the shifts between cold and warm phases in the Ice Age (Pleistocene), spanning around 100,000 years. We only observe higher values prior to the Pleistocene. Thus, the human-driven increase in atmospheric CO_2, when compared with values from geological history, already involves a geological timescale exceeding two and a half million years.

Projections suggest that CO_2 levels could exceed 1000 ppm in the coming centuries, resulting in an atmospheric CO_2 concentration similar to that of the early Tertiary or Mesozoic. Thus, in terms of atmospheric composition, the exploitation of fossil energy sources would encompass timescales of several tens to hundreds of millions of years. It is this extreme disruption of natural timescales that makes our current actions seem so problematic. The key problem is not the consumption of biologically generated energy storage, but the massive time scale violation in the use of the fossil resources (Fig. 3.6).

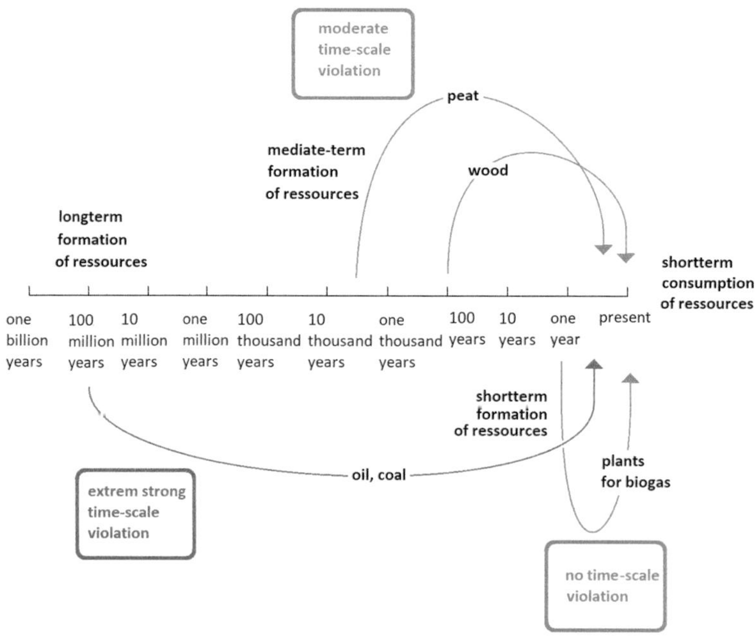

Figure 3.6 Massive time scale violation in the case of the use of fossil resources for energy production (logarithmic scale).

Entropy-export reserves accumulated in the form of fossil deposits over tens to hundreds of millions of years are now being depleted within decades to centuries. The ratio between the timescales of geological formation and consumption of these entropy-export reserves is thus approximately one million. This disconnects the timescales of formation and consumption of these reserves. The issue lies not in the mechanisms of fossil resource formation and use but in the disparity in the dynamics of these processes. This decoupling of timescales is the fundamental cause of the severe incompatibility of our current fossil-based energy production with Earth's natural system.

3.4 Time Scale Violation in Material Production and Use

Unfortunately, this ecological violation of timescales is not limited to the problem of energy production. The situation in the use of geological deposits for material extraction is not better. The material economy for our modern technology is currently organized largely without sustainability. Ores for metal production and many other inorganic raw materials predominantly come from deposits that originated in the Paleozoic, more than 240 million years ago. Construction materials are sourced partly from equally old resources, but also from more recent deposits such as limestone, chalk, and sandstone from the Mesozoic, or gravels and sands deposited in the Tertiary or Pleistocene epochs.

The long periods since the formation of the deposits currently being exploited indicate that we cannot, even in the long term, by human standards, rely on a natural regeneration of these resources. Natural recycling in the sense of sustainability is, under the current mode of raw material use, at best only feasible on a timescale of hundreds of millions of years. Additionally, not only has a long period passed between the formation of these deposits and today, but the natural formation of these deposits often took extended time periods as well.

Besides the consumption of long-term-formed natural raw material resources, the distribution of products and waste in the environment, and even the contamination of the biosphere, are serious problems of the use of matter from mines. The exploitation of ores, production, distribution of products, and release of waste and decomposition products into soils, waters, and the atmosphere represent a massive increase in entropy besides the consumption of energy, which relates to all production steps. This contribution to human-made entropy increase falls under the global ecological sin of time scale violation, too (Fig. 3.7).

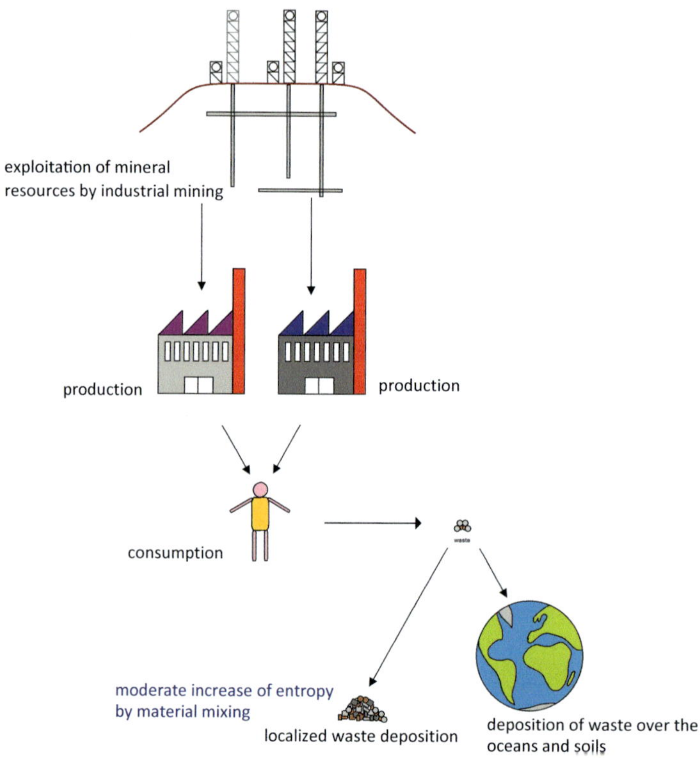

Figure 3.7 Time scale violation in the exploitation of natural raw material resources and distribution of products, side products, and waste in the environment.

The fact that some natural processes contributing to the formation of raw material deposits occur rapidly is of little consolation. These are generally short-lived catastrophic events that have created deposits: one might think, for instance, of the flooding of riparian forests and the burial of fallen and deposited wood under sands due to massive floods, or the formation of basalt deposits due to volcanic eruptions. The dramatic nature of these events underscores that they occur far from thermodynamic equilibrium. They are processes in which a pent-up thermodynamic potential is released in a short time. Such a release always means a rapid production of entropy.

Overall, the long geological timescales that we violate with our current resource-use strategies do not merely represent extensive time periods but also encompass dramatic shifts in environmental conditions, as well as rare yet potentially recurring regional and global catastrophes. A future sustainable economy must also include a sustainable approach to the outcomes of such catastrophes. Precautionary measures should aim to utilize stored natural thermodynamic potentials rather than allowing them to discharge uncontrollably with catastrophic consequences.

3.5 Comparison of Entropy Exports in Recent Industrial Processes with Early Human Societies

The industrial exploitation of natural resources for energy production and material extraction represents the entropy export of industrial society. Our society and economic system, too, are subject to thermodynamics. Therefore, they must be thermodynamically open systems. They need to produce entropy and must be capable of exporting more entropy than they generate so that previously achieved orders are preserved and new structures of order can be established.

The problematic global ecological impacts of industrialization have only come into public awareness in the past few decades. Industrialization itself began in tandem with the development of cities and the specialized production within them, a development that already began in antiquity. Even then, production and resource use were accompanied by human-induced ecological changes. However, it was only in the 19th century that industrialization reached global proportions. When looking at the whole history of humankind, it becomes clear that the industrial phase is a very recent phenomenon. Relative to the time since the emergence of humans about two million years ago, the industrial age is an extremely brief phase of development.

The look at the energy production shows that over the very long phases in prehistory, humans' energy needs were covered by sustainable resources, mainly wood. Only, in the younger human history, and namely since industrialization, fossil sources such as oil, coal, and gas have been exploited. That means the time scale violation in energy production is a comparatively young problem in human culture. In future, we must come back to a model closer to the situation in prehistory to overcome the massive time scale violation (Fig. 3.8). The key issue is the recently massively increased production of entropy and, therefore, the need of high entropy export from industrial processes and our highly energy consuming style of life.

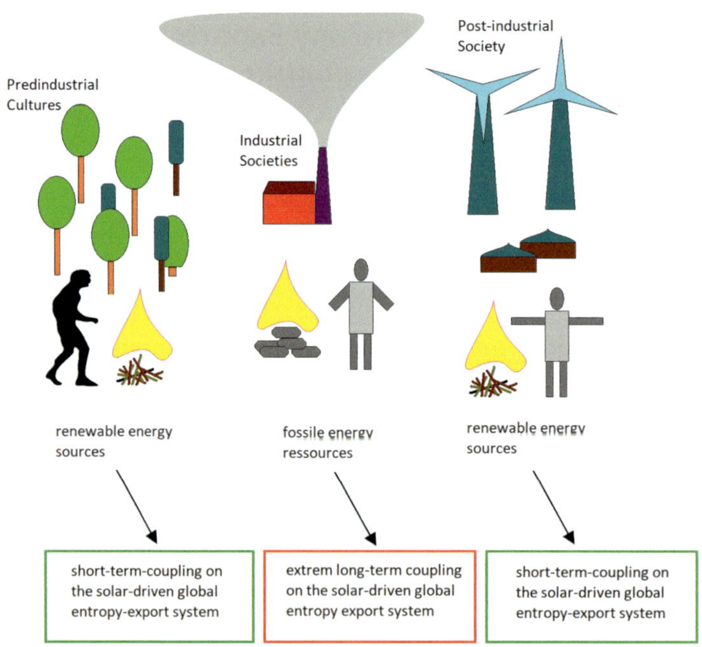

Figure 3.8 Comparison of energy production-related entropy export situation for prehistorical, industrial, and "post-industrial" societies.

For a comparison of raw material use between prehistory and the industrial era, it is useful to look at typical resource use in these periods. Over the long phases of pre-industrial human cultural development, renewable resources and energy carriers played the dominant role. Apart from stone tools, nearly all materials used for tools and everyday items before the metal age were renewable, primarily wood and bone. Energy needs were met almost exclusively by wood and other plant sources. Early societies required this consumption of natural resources, but these resources, along with the extraction of stone for tools and weapons, were sufficient to ensure the necessary entropy export. Resource use was largely limited to natural sources that could regenerate within decades to centuries. Their natural formation was directly and immediately connected to the constant energy flow from the Sun through the growth of plants and animals, supplying these resources. Thus, there was no significant violation of natural timescales. Early human societies created and used an entropy export system directly linked to the permanent entropy flow from the Earth to the cosmos, which is facilitated by the combination of solar radiation on the Earth and infrared radiation from the Earth. The only exception was the use of stones for tool production, which, given the vastness of the deposits and limited usage even over the long Stone Age, was considered unproblematic.

This situation fundamentally changed with the beginning of the Metal Age. The exploitation of copper and tin ores for bronze production introduced a resource that is comparatively limited in availability and whose formation, like that of many fossil fuel deposits, dates back hundreds of millions of years. However, the initial extraction was limited to ore-bearing layers emerging at the Earth's surface, and the impact of Bronze Age humans remained relatively minimal. The amounts and dynamics of material displacement associated with this use were still roughly comparable to natural erosion and displacement processes. But as near-surface ore extraction evolved into mining, the irreversible nature of extraction and material redistribution became more pronounced. Already in prehistoric times, especially since the Late Bronze Age, a significant amount of underground mining took place. Due to the locally limited scope of resource use, its global ecological impacts were not noticeable.

Changes became even more drastic with the use of machinery and the mechanization of mining processes. The massive scale of ore extraction during industrialization is exemplified by enormous open-pit mines and underground mining, where entire landscapes across hundreds or thousands of square kilometers were completely transformed, or geological layers were removed. In addition to the dramatic surface changes seen in open-pit mines, underground mining, though less visible, also caused extensive changes. The waste rock piles and tailings from resource processing represent only the visible tips of the iceberg. The broader changes occur within the geological layers of the subsurface.

A typical example of the progression of resource use from small-scale, localized mining to large-scale exploitation is provided by the copper shale mining in the Harz region and the Mansfeld region in Saxony-Anhalt (Germany). The use of copper shale likely began sporadically in areas of surface outcropping as early as prehistoric times, during the Late Neolithic or Early Bronze Age, around 4,000 years ago. With resource extraction and metallurgy conducted only in quantities ranging from kilograms to hundreds of pounds, the ecological impacts were very limited and initially negligible compared to the natural processes of weathering in geological layers.

The situation in ore mining in the Harz and other mountainous regions changed drastically with the introduction of underground mining in the High to Late Middle Ages. Initially, isolated shafts were created at particularly favorable locations, but soon hundreds of shafts lined the ore-bearing strata, from which ore was extracted. This ore was usually processed and smelted near the mines. The expansion of the mines and the processing of the ore required wood, and smelting led to an enormous increase in energy demand. Charcoal had to be used instead of wood to achieve the high temperatures required for smelting. This led to the extensive depletion of forests. In the service of metal production, humans transformed the landscape dramatically: where vast forests once existed in the Harz region, large barren areas began to emerge in the early modern period (16th–18th centuries).

This phase invites us to consider violations of natural timescales. The mining of ore already involved violating geological time scales. Copper shale was formed in the Zechstein, the youngest period of the Permian, which means it dates back about 250 million years to the late Paleozoic. For the ore, then, a significant violation of timescales had already occurred. The situation was quite different for energy provision. Although the widespread deforestation was a massive environmental intervention, it involved a renewable resource, wood, and even the charcoal used for smelting came from this renewable resource. Assuming a natural forest regeneration period of a few hundred years, the timescale here was roughly comparable to the technical processes of ore mining and metal production. Therefore, in this phase, there was not yet a severe violation of timescales concerning the resource for energy provision, in contrast to ore extraction.

The situation regarding energy use in copper shale mining fundamentally changed around the turn of the 18th to the 19th century. Forest reserves in both the immediate and even distant vicinity of mining and smelting areas were largely exhausted. Meanwhile, the demand for copper grew, the shafts became deeper, and mining outputs increased. The use of steam engines for ore extraction and, more critically, for pumping water from the mines necessitated more efficient energy sources than wood and charcoal. In the early 19th century, this need was increasingly met through the intensified use of coal deposits in the region. This shift in energy use led to a divergence between the timescales of resource utilization and natural regeneration in this area. Suddenly, an energy source was tapped that, like the ore, derived from deposits formed over extensive geological periods, rather than renewable sources that could regenerate within centuries. Thus, a profound violation of timescales became evident for energy provision in metallurgy as well.

The entropy production associated with ore extraction and metal production can be seen not only in resource depletion but also in the molecularly dispersed distribution of formerly concentrated raw materials. The carbon once concentrated in wood and coal is now distributed globally as carbon dioxide in the atmosphere. The copper initially concentrated as ore in deposits and subsequently extracted, and also undergoes a dispersion process. It has been extensively processed, exported, and distributed globally. Although it is sometimes reused in various items, copper ultimately ends up dispersed in landfills or as salts formed through corrosion in soils and waterways. With long-term deposition in the oceans, this provides a classic example of entropy increase through resource use and the dispersion of products derived from raw materials.

The destruction of natural reserves through metallurgy represents only part of the problem. As illustrated by copper extraction in the Mansfeld region, the invention of metal extraction sharply increased the demand for energy carriers for smelting ores. The overexploitation and destruction of forests observed in many areas are largely attributable to the enormous demand for wood and charcoal for smelting and metal processing. The rapidly growing demand for iron and steel was a primary driver of coal mining during industrialization, and a substantial proportion of fossil fuels is still used today in the form of coal and coke for metal production.

The development of metal ore mining, metallurgy, and metal-based production was connected with an increase in exploitation and overstretching of natural resources. This development, including the transition from copper and bronze metallurgy to iron production, and further iron-based recent industry, was accompanied by an increase in the problem of time scale violation (Fig. 3.9).

In addition to the unsustainable use of fossil fuels, the current use of mineral resources is also unsustainable. Besides copper and iron ores and other metal deposits, construction materials represent significant raw materials that require large quantities and lead to intensive environmental interventions. Gravel and sand are needed in vast quantities worldwide. Their deposits are sometimes naturally renewed through erosion, marine dynamics, glaciers, and rivers relatively rapidly by geological standards. Nonetheless, large amounts of materials are being extracted from deposits that are millions of years old. The movement of

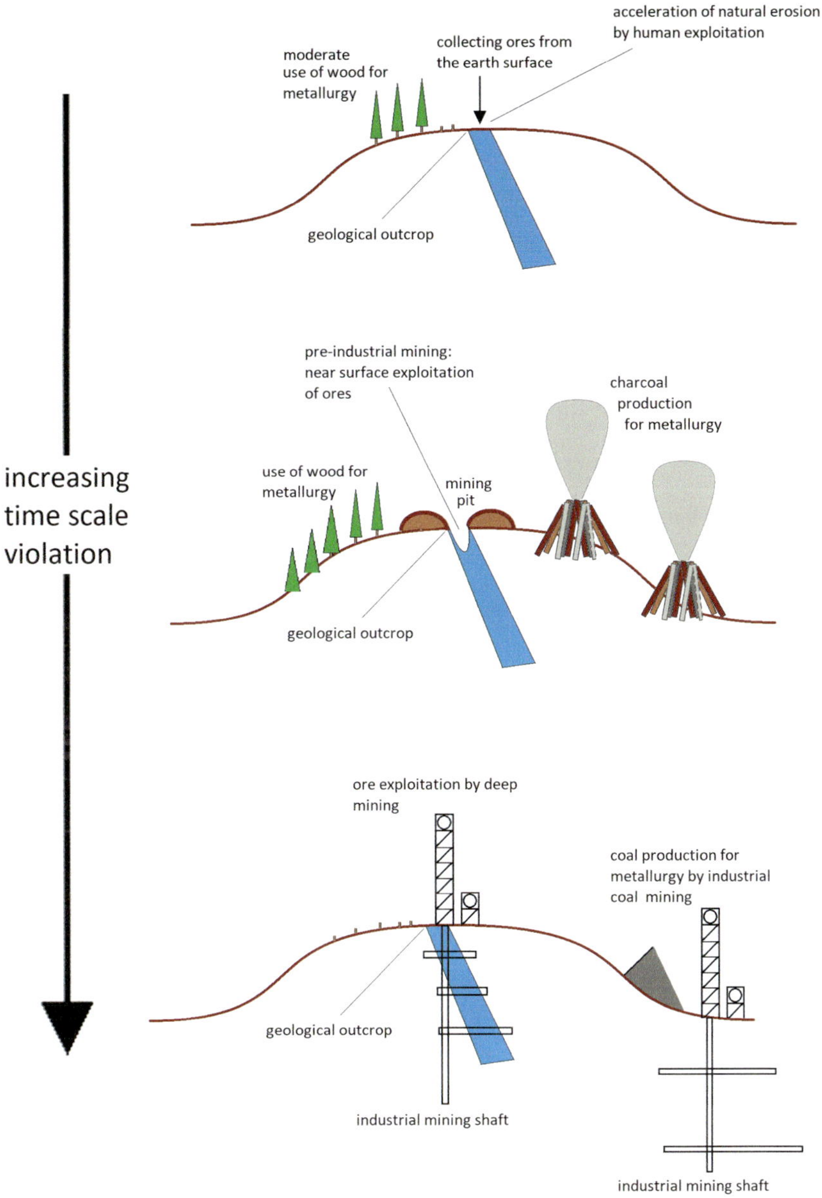

Figure 3.9 Increasing strength of the time scale violation problem in the development of metal-based industrial society.

water and air during sediment formation has often determined that the sedimentation of solid materials was influenced by the density and size of the solid particles. Classified by material and grain size, these deposits represent large, ordered units and interesting reserves for entropy export. They are products of atmospheric and surface processes driven by water movements, atmospheric processes, and indirectly by solar energy, thousands or even millions of years ago. In this sense, they also represent negentropy derived from the former flow of solar energy and the degradation of high-energy solar photons to long-wave photons emitted by the Earth into space.

Similarly, this applies to mineral resources used or processed as fertilizers in agriculture. Worldwide, phosphate deposits, essential for both industry and agriculture, are becoming scarce. Potash salt deposits, also required in large quantities as fertilizer, and lithium deposits critical for battery production are generally products of the evaporation of lakes or seas, which are ultimately processes driven by solar energy.

Geological deposits, regardless of type, always form such "negentropy stores" created in past eras. They are an expression of Earth's entropy export into space and, at the same time, the result of Earth's ability to use global energy flow to build structures and thus order. Each deposit exemplifies the spontaneous formation of such structures. Every natural process of material concentration or complete separation, each classification by grain size, is an order-forming process that relies on a coupling to an Earth-bound or cosmic entropy export mechanism. Geological structures, including those not considered for resource extraction today, should be counted among these "negentropy stores."

The transition from pre-industrial to industrial production involved a shift from moderate to drastic timescale violations, not only regarding energy production but also in material management. In extracting construction and industrial materials, the young industrial culture's demand for entropy export capacity soon exceeded what sustainable natural processes on the Earth could provide. The shift from wood to coal and to largely non-renewable fossil fuels is a typical consequence of this heightened demand for entropy export capacity, as is the drastic increase in demand for building materials.

The speed of this shift can be vividly illustrated by comparing topographical maps. Between the mid and late 19th century, European cities suddenly expanded drastically. Industrialization brought a sudden surge in spatial requirements. Cities burst from the narrow confines they had been limited to for centuries by city walls. While mid-19th-century maps show most cities as compact, small-scale structures, by 1900, these cities had become sprawling settlements and industrial areas, extending well into surrounding landscapes. This trend continued in the 20th century as ever-larger swathes of land were taken over by cities, residential and commercial areas, and transportation infrastructure, increasingly restricting agricultural and forestry use.

The rapid expansion of settlement, transportation, and industrial areas naturally drove an increasing demand for building materials. Thus, urban growth also reflects a rapid depletion of usable raw material deposits. By tapping geological deposits, humanity draws on nature's long-term negentropy reserves to meet its demand for entropy export capacity, exceeding what natural processes can sustain.

This shift from quickly renewable natural resources to geological sources for energy and materials is, in every respect, a result of heightened entropy production. Over the past two centuries, the mechanisms of industrial economy and society have led to ever-increasing global entropy production, accelerating the consumption and exhaustion of global entropy export reserves formed over geological timescales.

The role of entropy export in creating mineral order structures through the energy flow from the sun, via the Earth, to cosmic space is particularly clear in the example of limestone deposits. Many mountain ranges consist largely of calcium carbonate or calcium magnesium carbonate (dolomite), representing typical examples of concentrated materials dominated by a single chemical substance. In some limestone deposits, calcium carbonate even occurs in a sufficiently pure form to be directly usable in high-purity industrial processes, such as in the production of optical glass.

Polymers derived from coal and oil now substitute natural fibers in textiles and ropes and replace construction materials in architecture and engineering. The unsustainable exploitation of deposits, the inability to renew them, and the ecological consequences of production and distribution have all reached global dimensions.

3.6 The Problem of Small-Scale-Structured Designs of Materials and Devices

In the course of technological culture, not only does the volume of unsustainably sourced and used raw materials increase, but alongside the growing tonnages of production, the proportion of small-scale products also rises, with many products becoming ever smaller, down to micro- and nanometer dimensions. Different materials are being combined in increasingly complex and diverse ways to achieve specific material properties and functions.[5] And here, too, the resulting material structures support cultural order structures, at least in the short term. In the long term, however, most of these small-scale products become waste that must be laboriously processed or disposed of.[6] Here as well, the violation of time scales is the central problem: the materials for small devices, such as household appliances, vehicles, mobile phones, and computers, largely come from deposits formed over tens of millions of years, yet their functional lifespan is at best measured in decades, often only a few years. The smallness of this waste micro-objects represents a new level of disorder.[7] The globalization of this disorder, along with the unchecked production and distribution of an ever-growing number of short-lived products, indeed represents a pattern in new social structures and is therefore part of social order, but at the same time creates an increasingly severe global ecological disorder (Fig. 3.10).

[5] Y. Yu et al., 2000.

[6] B. Adám et al., 2021.

[7] F. Tommasi and L. Mancini, 2019.

The smallness of many modern products reflects the ambivalent nature of information and structural richness. As functional objects, modern devices represent accumulated information and thus negentropy. Their components have been arranged in a specific spatial order to fulfill certain functions. These devices are part of the system-inherent knowledge within technological culture, and therefore within human society. However, after their use has ended, they lose this function and become waste. The original information becomes worthless. All that remains is a high level of disorder due to the distribution and smallness of the discarded products. The once functional structure has become an increase in entropy.

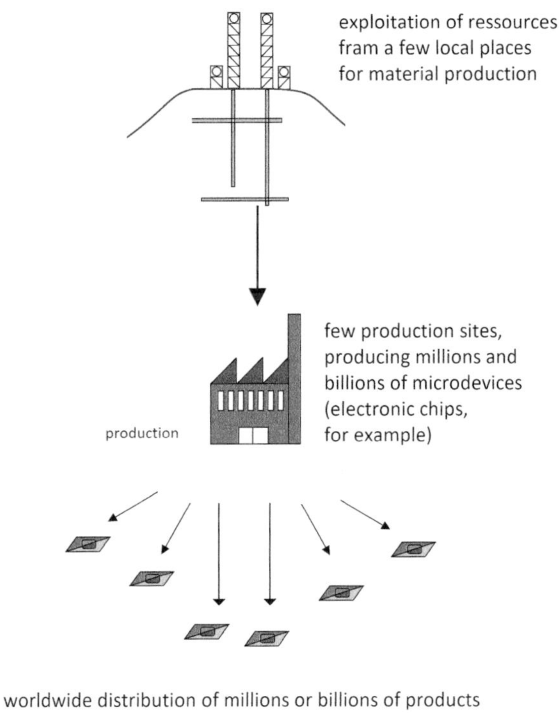

Figure 3.10 The production and distribution of micro- and nanostructured devices represent a special form of entropy.

The recyclability of small-scale components and devices can be improved if even the smallest separate elements carry information that supports their recognition, collection, and further processing. Such information and matching read-out systems could essentially promote the recycling processes.[8]

3.7 Possible Potential of Small-Scale-Structured Design of Materials and Devices

Perhaps, however, the trend toward the miniaturization of products also presents an opportunity. The concept of miniaturization could offer an approach to initiate a shift in material use without requiring the abandonment of complex and widely used industrial products. Miniaturization in industrial products has a dual nature. On the one hand, it involves the depletion of deposits of concentrated materials, i.e., order or negentropy, and the mixing and dispersion of different materials, which increases disorder, leading to an accumulation of entropy. On the other hand, miniaturization represents an approach to the biological model of living systems, as they two are fundamentally organized in a miniaturized way regarding the components and materials used.

In essence, there is a convergence between the miniaturization of complex living systems and modern industrial products in terms of their fundamental structure. On the one hand, it is the complexity of functions that necessitates a hierarchical organization. Structure and function represent system-specific information, which requires an entropy-generating process coupled with an entropy-export mechanism for its creation. Structural miniaturization is a prerequisite for complex functionality. It is no coincidence that the smallest components dedicated to information acquisition and processing, such as sensors and electronic processors in modern technical systems, are also the most miniaturized and intricately structured. On the other hand, a hierarchically organized structure built upon small elemental components is an essential principle for establishing spatial order across several orders of magnitude in length and volume (Fig. 3.11).

[8]K. Mishima and H. Nishimura, 2016.

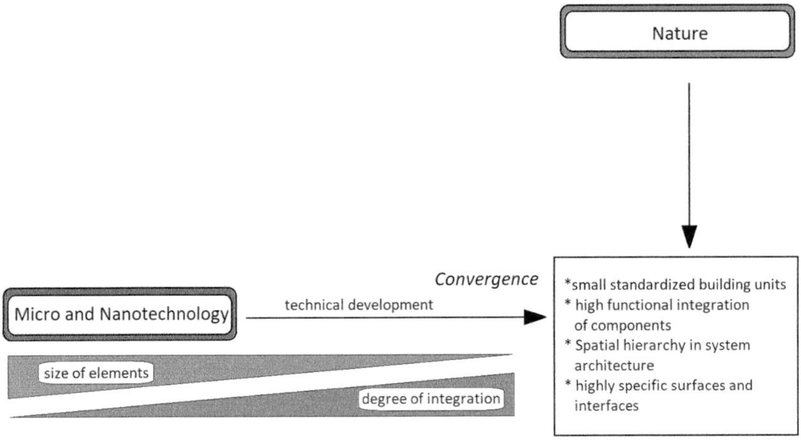

Figure 3.11 Analogies between the construction and production of micro- and nano-scaled devices and living nature support the idea of achieving sustainability by a further convergence of technical and natural processes and components.

In nature, miniaturization enables efficient integration into material cycling. Miniaturization is generally associated with large boundary and surface areas. Degradation and breakdown processes, which involve interfacial reactions, typically occur more rapidly in small components than in compact objects. Thus, miniaturized systems are more sensitive, more influenced by their surroundings, and more repairable, regenerable, or degradable. This is especially true for porous, sponge-like, gel-like, or porogenic materials, in which porous structures can develop from a compact state. Porous and porogenic structures and various hybrids are found in abundance both in nature and in technology. Examples include sponges and corals, wood, and bone, but also ceramics, polymer composites, textiles, solid foams, and aerogels.

However, miniaturization does not inherently imply "similarity to nature." Even less does it guarantee inherent environmental compatibility. From a thermodynamic perspective, miniaturization represents a state of increased order as long as a specific purpose or systemic function is fulfilled by the pattern of miniaturization. When this function is lost, e.g., when a living being dies, the value of the order is lost. Especially when an industrial product no longer functions or is no longer needed and must be disposed of, miniaturization can often be a significant obstacle to recycling.

Ultimately, it is not miniaturization that determines ecological compatibility but the integration of material regeneration processes with natural entropy-export mechanisms. For this purpose, not only the structure but also the elemental composition and the compatibility of degradation and transport processes with natural cycles are important for reintegrating components into natural recycling processes.

Chemical elements and compounds that are abundant in the mobile parts of our environment and frequently cycled through sun-driven mechanisms, thus exhibiting high natural cycling rates, are ideal for the creation, regeneration, and decomposition of systems that aim to be eco-compatible. Elements like oxygen, hydrogen, and carbon, and simple compounds such as carbon dioxide and water fulfill this requirement ideally. In a similar sense, the broad range of low-molecular-weight hydrocarbons and their derivatives is also suitable. While these may individually often be toxic or contribute to climate change, they are compounds that, in their formation and degradation, meet the criteria for good potential integration with natural chemical and biological processes. Indeed, it is this very compatibility with natural cycles that sometimes contributes to their negative biological effects. Considering the abundance of these elements and their low-energy compounds in our waters and atmosphere, they represent ideal building blocks. Often, we rely on ancient reserves of coal, oil, and gas instead of renewable carbon sources simply because the former require less energy input and are more cost-effective. Beyond this, the central role of elements like carbon, oxygen, and hydrogen in living systems is not only due to their favorable chemical properties but also to their universal availability and the possibility of their natural processing within global material cycles driven by solar energy.

In contrast, incorporating rare elements and compounds that are only minimally cycled in nature inherently involves longer cycle times and a less efficient restoration of original natural conditions. Elements like heavy metals (copper, cadmium, mercury, and lead) or rare but less toxic metals from the platinum group (palladium, rhodium, ruthenium) are far more challenging to integrate into natural cycles. Extraction from deposits where these elements are present in significantly higher concentrations than in the Earth's crust average composition typically results in their eventual dispersal in soils and waters, ultimately reaching the oceans. This transition, from minimal entropy in concentrated deposits to maximal entropy when these elements spread through wastewater into the seas or are dispersed as fine dust, is a movement from order to disorder.

Living nature does use rare elements in cells, albeit sparingly. Their use in nature is often due to specific chemical, particularly catalytic, properties of certain heavy metals. Unlike industrial catalysts, which are used in high concentrations or as compact materials in plates, nets, or large clusters, biocatalysts containing metals are mostly composed of carbon, hydrogen, nitrogen, and oxygen atoms, with thousands of these atoms typically surrounding a single metal ion. This arrangement maximizes the metal ion's unique electronic properties and catalytic efficiency, thus enabling an efficient use of the rare element. The living cell invests significant effort in structuring the surroundings of the metal ion to optimize its catalytic function within a metalloenzyme.

Thus, in the elemental composition of living systems, one generally observes a trend toward aligning the ratios of elements within the systems to those of their environmental availability. A similar trend of adapting availability to usage can be observed in technological history. Notable examples include the early replacement of copper and bronze with iron and steel for tools and weapons, or the switch from germanium to silicon in semiconductor technology. In both cases, humanity benefited from the superior functional properties of the newer material, iron and silicon, which not only rank among the most abundant elements in the Earth's crust but also offered superior functional properties for widespread technological use.

Typical semiconductor components, like silicon-based ones, are much more miniaturized than average steel tools or device components. However, even more miniaturized than silicon are thin-film functional materials—these include not only silicon dioxide and silicon nitride but also microstructures made from rare elements, such as gold for conductive paths or compound semiconductors. With ongoing downscaling in semiconductor technology, we have seen a decades-long trend toward smaller dimensions and increased integration density (Moore's Law), resulting in progressive miniaturization paired with growing complexity. Ultimately, molecular- or cluster-based components could be the future, bringing miniaturization in technical systems down to the atomic scale, like what we see in living nature.

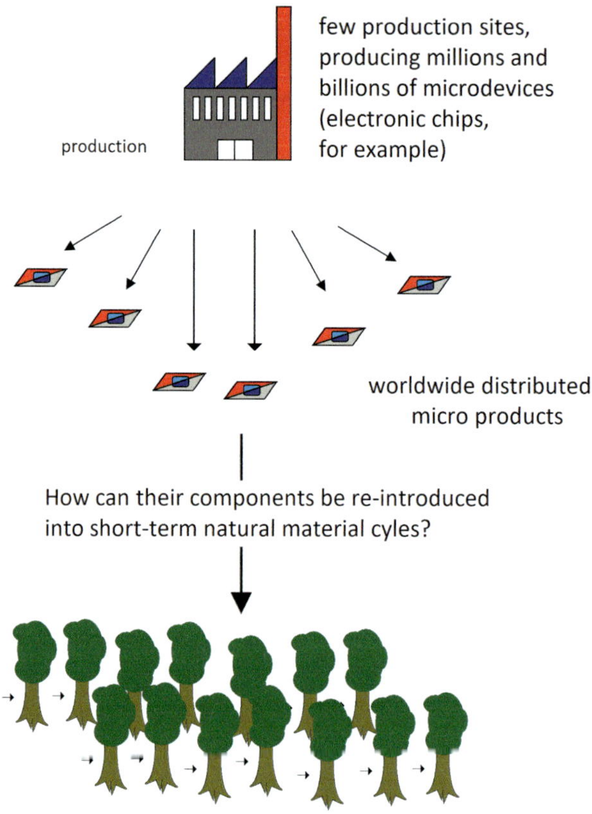

Figure 3.12 Vision of integration of industrial micro and nanoproducts into short-term natural recycling processes.

But to be honest, we must recognize that the used materials and recently produced devices, especially in the case of the semiconductor industry, are due to the technical requirements, are very far from biomolecular systems in living beings. Thus, it will probably still be a long way before we can produce and apply micro and nano products which allow us to use the advantages of convergence between industry and nature for a complete integration of technical products in the short-term natural material recycling (Fig. 3.12). Besides the problem of material composition, the problem of adaptation of technology must also be solved.

Interestingly, the convergence between the structural complexity and miniaturization of technical and natural systems can also be observed in their assembly principles. While traditional technology assembles complex systems by mounting functional units composed of smaller functional subunits, semiconductor technology forms transistors, capacitors, and other elemental electronic units concurrently with the integrated circuit. This means components and systems are created not sequentially but simultaneously, a technological principle mirrored by the development of a complex organism from a small initial state, the germ, as it simultaneously forms its tissues, organs, and overall system.

3.8 Need for Time Scale-Adaptation

The example of highly integrated solid-state circuits strikingly illustrates that small-scale design does not automatically enable rapid reintegration into natural material cycles. As long as natural recycling relies on processes that already occur in nature, one would likely have to wait for geological processes to return these materials to a usable state. Only when landfills containing such components and materials are subjected to sedimentation and tectonic forces, pushing them into deeper regions of the Earth's crust and possibly the upper mantle under high pressure and temperature, could natural "processing" begin through melting, geochemical reactions, and transport. This would mean reverting to the geological time scale.

The natural global ecosystem also provides us with examples of extremely long-lasting material cycles. Whenever the geological movement of plate tectonics is the slowest step in a cycle, the time scales extend into hundreds of millions of years. This is the case in cycles where sediment layers are built up and only undergo significant thermal impact and conversion into volatile products through lateral plate movement and subsequent subduction into the Earth's mantle. The classic example of this is the natural release of carbon from sediments in the form of carbon dioxide through the thermal decomposition of carbonate rocks. Such sediments are generally of biogenic origin and are only reintroduced into new chemical processes when forced down to zones in the mantle that are sufficiently deep and hot. Inevitably, the time intervals between the biogenic formation, sedimentation of carbonates, and the endogenous thermal decomposition of these layers match the time spans of continental drift, ocean formation, and disappearance. Due to their link to geological dynamics, the natural life spans of ore and salt deposits are similarly on the same scale.

The concept of sustainable resource management, which we must achieve in the short- to mid-term, is therefore focused on cycles that are directly linked to biological growth and the rapid transport of materials through the water cycle and atmospheric transport systems. All three cycles are continuously driven by solar radiation and are thus directly connected to the global entropy export mechanism. Withdrawing materials from these cycles and returning waste to them means aligning with their natural time scales. This fact provides the primary argument for using renewable resources. We must avoid that products and byproducts are formed which cannot be integrated in the short-term natural conversion processes and must be deposited (Fig. 3.13a). Instead, we need closed material cycles, which means that all paths of matter have to be connected with the natural short-term processes (Fig. 3.13b). This corresponds to the fact that man and technique must be completely embedded in natural networks, in biological and metabolic processes.[9]

Against this backdrop, it is initially easy to identify materials that cannot be derived from renewable resources. In addition to iron, copper, and all other industrial metals, this includes

[9]L. Pellizzoni and L. Centemeri, 2022.

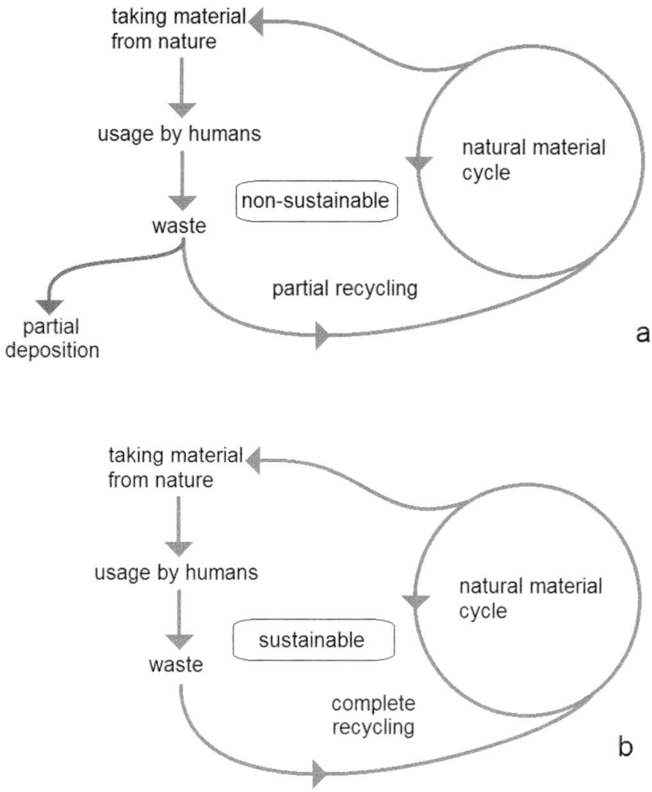

Figure 3.13 Coupling of materials use with natural cycles: (a) Partial recycling means non-sustainable material management, and (b) Only complete recycling means sustainability.

mineral-based building materials and most synthetic polymers. Given the widespread and, in many cases, indispensable use of these materials, it is much harder to envision a complete or even partial substitution with renewable resources. To address this challenge, three groups of problems must be solved. First, substitutes must be found, based on sustainably sourced materials, to fulfill the functions of today's industrial materials and consumer products. Second, sustainable processes must be developed to produce these substitutes in sufficient quantities (mass production) from renewable resources. And third, these materials must be converted at the end of their functional life so that they can be reintroduced into natural cycles without loss (Fig. 3.14).

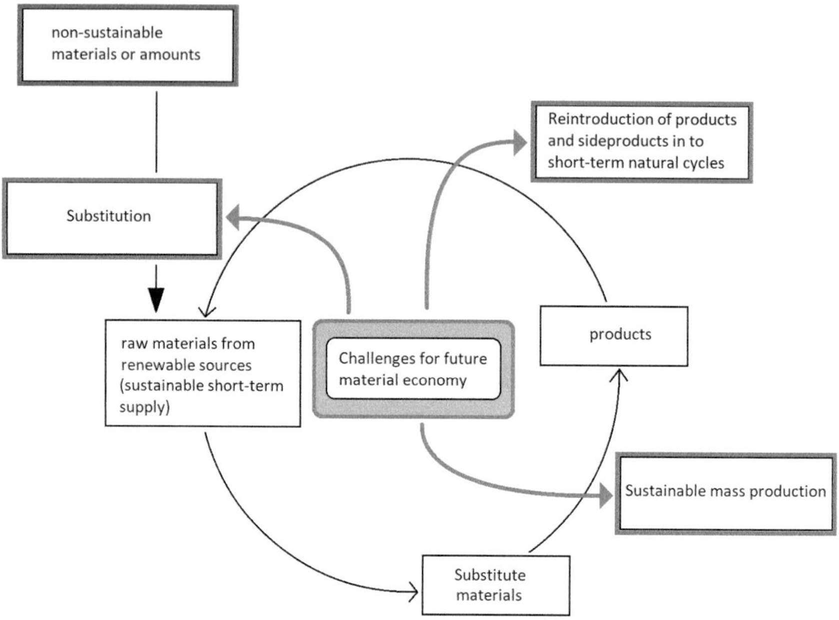

Figure 3.14 Challenges for future sustainable materials, product management, organizing a closed cycle between raw materials, converted materials, and products.

In the coming decades, the entire materials industry will need to transition from its current largely unsustainable model to a sustainable one. The recent know-how in processing of bio-based raw materials and in particular the recently already available technology for food production can help to develop the required technologies for new bio-based materials using in a large spectrum of applications.[10] Managing the energy required for material use and the entropic aspects of material utilization and transformation will play a central role.[11] The "entropy transition" can only be achieved by linking the "energy transition" with a "materials transition."

[10]M. Loeffler et al., 2018.

[11]For example, by cellulose-based materials (J. Verma et al., 2024).

3.9 Sustainable Entropy Management in Agriculture

Since its inception, agriculture has held fundamental economic and political significance due to its central role in human nutrition. In addition to food production, agriculture has always provided plant materials for a wide range of purposes. To this day, this includes agricultural products for energy production, textiles, construction materials, and many other uses. In principle, agriculture has also played a large ecological role since its beginnings. However, in recent decades, it has become clear that this ecological significance is not only local but also has strong global impacts.

Agriculture, livestock farming, and forestry all follow the core principle of production through the use of plant photosynthesis. From a thermodynamic perspective, these practices harness the potential created by the photonic entropy gradient between visible sunlight and Earth's thermal radiation to produce energy-rich materials from low-energy raw materials, primarily carbon dioxide and water, that can be used as food, luxury goods, feed, active ingredients, energy sources, construction materials, and other resources. In thermodynamic terms, the function of green plants is often reduced to energy transformation and the associated export of entropy. Photosynthesis, however, also fulfills a second role: it accumulates dispersed substances and generates short-term concentrated material reserves and long-term storable resources. Thus, agriculture and forestry also serve a dual purpose. They provide substances and energy carriers and reduce global entropy, characterized by the dispersion of substances, through the fixation and conversion of mobile small molecules and the concentration of produced substances. In this way, agriculture and forestry are essential global tools for the long-term reduction of atmospheric CO_2 levels. Additionally, using significant portions of the Earth's surface for photosynthesis-driven accumulation enables the fixation and storage of other dispersed components, thereby reducing their contribution to global entropy and potentially making them available for technical use. This includes dispersed metals, organic environmental pollutants, and nanoparticles, which are often environmental toxins in distributed form but become valuable resource sources when accumulated.

The reliance on sunlight makes agriculture and forestry inherently sustainable at their core, yet it also necessarily ties them to land use. In no other economic sector does land use have such central importance as it does in agriculture and forestry. Indeed, one could say that the effective use of significant portions of the Earth's surface for production via sunlight, coupled with the reduction of global entropy caused by material dispersion, is the defining characteristic of agriculture and forestry. This connection underscores the unique role of these sectors. Future factories that, for instance, produce fuel from atmospheric carbon dioxide will require an external energy supply. Photovoltaic fields and solar thermal power plants, which provide energy by accumulating and converting solar energy on large surfaces, do not contribute to material fixation. The large-scale integration of solar energy use and material transformation has, to date, only been achieved in agriculture. Therefore, agriculture and forestry assume a central position in realizing the "entropy transition" (Fig. 3.15).

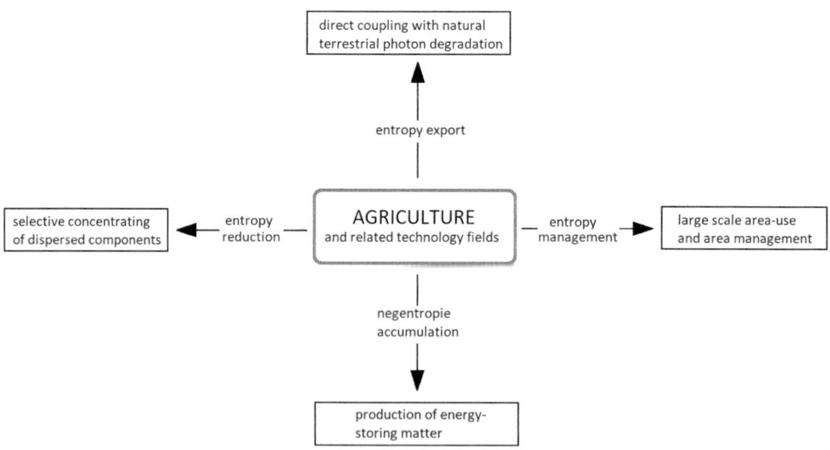

Figure 3.15 Central position of agriculture, forestry, and related technology fields in future sustainable entropy management.

In addition to the direct linkage of material transformation and solar energy use, agriculture and forestry hold another fundamental advantage over industrial contributions to the entropy transition. The systems employed—green plants and their ecological surroundings, including animals and microorganisms—are ecologically interconnected and thus integrated into natural material cycles, linked with the natural flow of biological information. This form of production has high inherent bio- and eco-compatibility and can leverage the entire adaptive and evolutionary potential of biological systems, including modern biotechnological methods derived from them. These fundamental characteristics of biologically driven land use are likely to become crucial in achieving the entropy transition. Consequently, a future economy based on comprehensive sustainable entropy management must be largely supported by the cultivation of the Earth's surface, with agriculture and forestry playing a decisive role.

Chapter 4

Entropy Export in Sustainable Material and Land Resource Management

4.1 Role of Agriculture for Sustainable Material Production and Cycling

The modern global materials economy is largely shaped by the use of geological deposits and mineral-based materials, a direct consequence of industrialization. Over recent centuries, industrialization has replaced organic materials, primarily derived from biogenic synthesis, with more efficient alternatives that are often faster and cheaper to produce. Synthetic textile fibers are a typical example. Before the invention of synthetic polymers, textiles were made from plant or animal fibers. Natural raw materials sourced from short-term natural cycles, which could easily return to these cycles after use. With the advent of synthetic polymers and fibers, natural fibers were largely replaced by those derived chemically from coal or petroleum, tapping into fossil resources that, like fossil fuels for energy, are millions of years old. The shift in textile fiber production from natural to synthetic materials has, therefore, drawn textile production into the significant timescale violation inherent to industrial production.

Entropy and Sustainability: Challenges of Future Energy and Material Management from a Thermodynamic Point of View
Michael Köhler

ISBN 978-981-5129-99-1 (Hardcover), 978-1-003-74889-2 (eBook)
www.jennystanford.com

DOI: 10.1201/9781003748892-5

Moreover, industrialization led to the creation of tools and machines that could only be manufactured from mineral-based raw materials, such as metals. This development inevitably imparted the predominantly non-sustainable character of the industrial materials economy that persists today. This lack of sustainability impacts not only metals but also oxide-based and salt-like mineral resources, which are extensively used in industries like construction, as well as synthetic organic materials like plastics.

A fundamental shift in this situation is needed as part of a "materials transition." This requires a significant realignment of agriculture and forestry toward supplying biogenic materials, alongside a radical conversion in materials used in today's technology. Besides the material production and use itself, the "material transition" must include a strict connection with the sustainable production of all needed forms of energy, and preferably an input of needed carbon into the technical processes by accumulation and reduction of CO_2 from the atmosphere. The synthesis of technical primary products in the form of basic organic chemicals should use sunlight directly (photosynthesis) or sustainably produced electrical power (Fig. 4.1).

Replacing individual metal components in fields like automotive and aerospace manufacturing with plastic elements, or supplementing inorganic solid-state electronics with organic semiconductor electronics, represents progress but falls far short of the necessary final state. The scale of the challenge can be gauged by imagining engines, chemical plants, and aircraft turbines made from agriculturally sourced materials (Fig. 4.2). For complex technical solutions with a relatively long development history, there will be no rapid material conversion. Interim solutions and decades-long development and implementation periods will likely be required. For this reason, efforts to completely substitute non-sustainably produced materials with biogenic materials must be intensified and expanded, even into areas where particularly high chemical, mechanical, and thermal durability of materials is required.

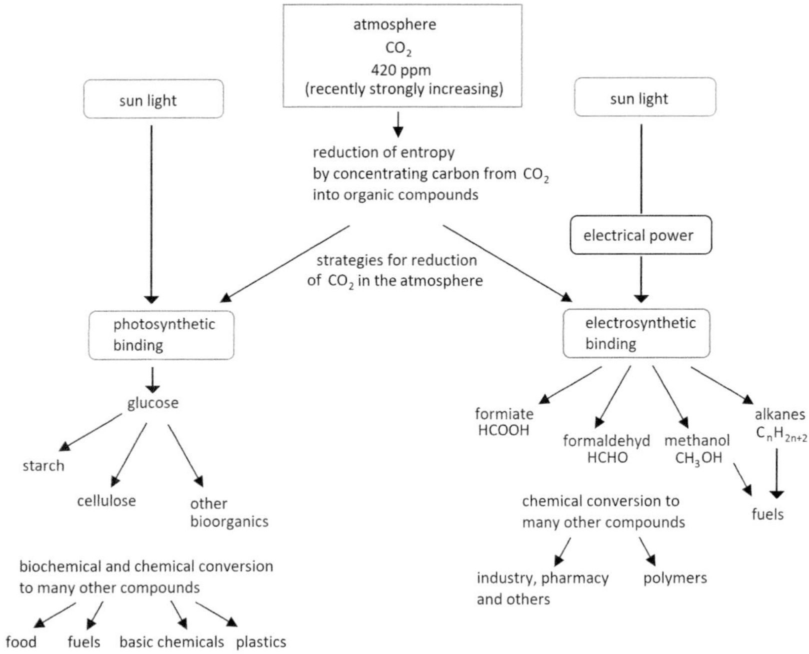

Figure 4.1 Accumulation of carbon from the atmosphere for the sustainable production of basic organic compounds using energy from sustainable sources.

These challenges can only be addressed through a complementary reorientation of technical material applications, design, and industrial technologies on the one hand, and agricultural and forestry production, including the plants and biological communities employed, on the other. This will also necessitate fundamentally new technologies in agriculture and forestry. It is expected that the currently distinct boundaries between agriculture and forestry will fade, as the core principles of these new technologies must be applied in both areas. This includes aspects such as perenniality,[1] mixed cultures, and locally selective cultivation measures, as well as the use of genetically modified crops and co-culture components, extending even to the design and cultivation of small organisms and microbial communities.

[1]Enhanced crop diversity and perenniality improve soil properties (C.D. Sprunger et al., 2020).

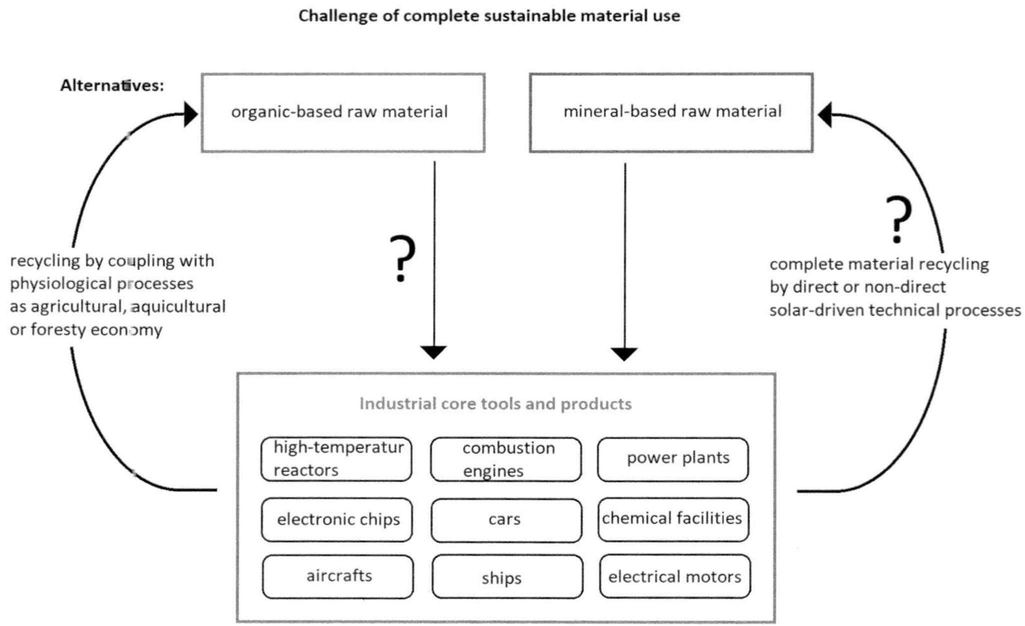

Figure 4.2 Alternative strategies for the material basis for sustainable production of complex technical systems: replacement of conventional inorganic materials by organic substances (green, left) or complete integration of inorganic materials and devices produced from them into nature-coupled recycling processes (red, right).

To meet the new, comprehensive requirements of the global economy with a fully sustainable entropy management, the current, more traditional forms of agriculture and forestry will likely be replaced by a "biosolar economy." This term could encompass all activities related to production, disposal, and recycling supported by photosynthesis, whether they pertain to food production, the provision of biogenic fuels, or the sustainable production of raw materials for technical applications.

4.2 Biosolar Economy

The concept of a "biosolar economy" outlines a global entropy management system entirely based on the flow of light energy from the sun and the emission of heat radiation, in which biological photosynthesis plays a central role. Industrial-technical energy conversion processes are included in this framework only to the extent that they involve both material and energy transformation—similar to biological photosynthesis—that reduces entropy related to the distribution of materials on the Earth, and are achievable through facilities and methods that fit into a fully sustainable global material cycle. Accordingly, wind and photovoltaic systems are only suitable for a biosolar economy if the energy they provide is directly or indirectly usable for the fixation and deposition of dispersed substances, such as atmospheric CO_2, and if the systems are entirely made from biogenic materials that can fully break down into biologically useful components after their use. As such, today's existing and developing systems serving the energy transition should be viewed as essential elements of bridge technologies, which will need to be replaced in the foreseeable future by systems fully integrated into a completely sustainable economy (Fig. 4.3).

Why must a fully sustainable future economy be based on biological systems? Is it not conceivable to create technical systems that are entirely recyclable? Would it not suffice to base energy production entirely on entropy export through photon degradation directed by sunlight into the cosmic background, i.e., converting visible and UV sunlight into infrared radiation?

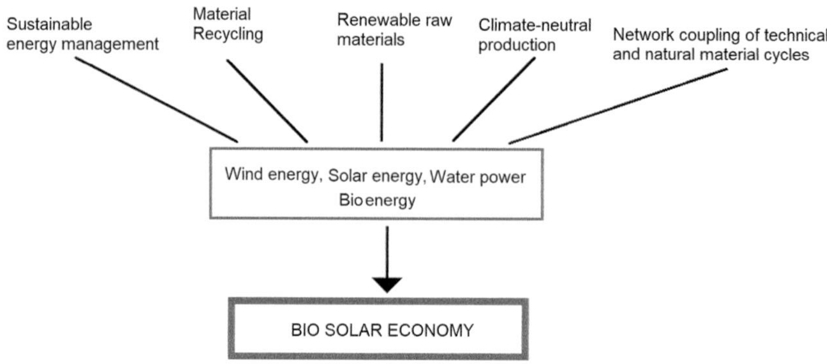

Figure 4.3 General components of a Biosolar Economy with short-term coupling between technical and natural material cycles and energy supply from renewable sources.

In principle, this role could also be fulfilled by technical systems. The extensive photovoltaic and wind power fields being established globally are current examples of this. Such systems are likely to play a central role for decades, perhaps centuries, in transitioning to a sustainable energy economy. Nevertheless, it will most likely remain a bridge technology since, while the energy production is sustainable, the material components still include mineral-based, non-sustainable materials, which means the system does not align with a fully sustainable economy. All functionally necessary components of energy systems would either need to come from biological production or be sourced from seawater under economically viable conditions. This requirement for seawater sourcing is based on the notion that, over long periods, most components of technical waste will eventually reach the sea in a corroded form, with most metals ending up as salts or oxide particles. The dispersion of waste materials into the environment and the oceans is a typical process driven by the entropy of distribution. As trace elements in seawater, chemical elements reach their highest possible entropy, thus requiring a large amount of energy for metal recovery from seawater to achieve the entropy export necessary for concentration.

If metals and metalloids were to be avoided, photovoltaic solar cells, for example, would need to be constructed entirely from organic materials derived from biological production, including semiconductors and metal-based wiring. It is also difficult to imagine that the windings of wind turbine generators could be made from organic conductors. If, however, sustainable metal extraction is accepted, the key metallic components of today's systems would need to be enriched from seawater, for example, through "metal farming," to meet sustainability criteria. This would only be economically viable if the systems operated for long periods and underwent many recycling cycles before residues ultimately reached the oceans.

The central importance of biological material provision, which relies on natural photosynthesis, stems from the need for a fully sustainable entropy export. An economy that derives all its material needs from photosynthesis has the best prerequisites for linking both production and disposal directly to natural biological cycles.[2] Besides plants, photosynthetically active microorganisms can be used for the generation of bio-based raw materials.[3]

Material generation would occur in complete analogy to natural assimilation. In return, surplus organic materials from products could be biologically degraded or reintroduced into biological cycles. The technical materials economy would thus become an integral part of the natural global metabolic process and global matter exchange and transformation. Most metals, especially rare ones, are needed by biological processes only in trace amounts. For heavy metal ions, which act as central atoms in many enzymes, suitable biochemical mechanisms already exist in living organisms to extract them from the natural environment. The required amounts are so small that living nature can afford the energetic cost, i.e., the entropy export required for obtaining them despite very low environmental concentrations.

For other specialized technical elements requiring significant amounts of metals and semiconductors, achieving sustainability is more complex. Their production and disposal would require technical process chains in which all steps fulfill sustainability criteria. Such an investment would be justifiable only in rare

---

[2]Razeghifard, 2013.

[3]R.U. Tanvir et al., 2021.

cases for essential functions. In all other areas, photosynthesis-based material provision and a production and disposal system fully coupled to biological cycles will dominate. As with today's agriculture and forestry, this bio-based energy production will involve area-dependent bioproduction driven by sunlight. Agriculture, forestry, so-called "agroforestry,"[4] and a sunlit, expansive "aquaculture economy"—distinct from vertical, artificially lit aquaculture—will form the foundation of this biosolar economy, which will underpin not only food and feed production but also the provision of materials and energy. Together, agriculture, forestry, agroforestry, and sun-powered aquaculture are termed the "biosolar area economy" (Fig. 4.4). The shared, defining feature of these sectors is the stewardship of the global habitat to achieve efficient, fully sustainable conversion of solar energy into various forms of biomass. The biosolar area economy combines the use of large portions of Earth's surface to sustainably supply materials and energy with the complete recycling of by-products and end-products, thereby maintaining a vital and resilient biosphere.

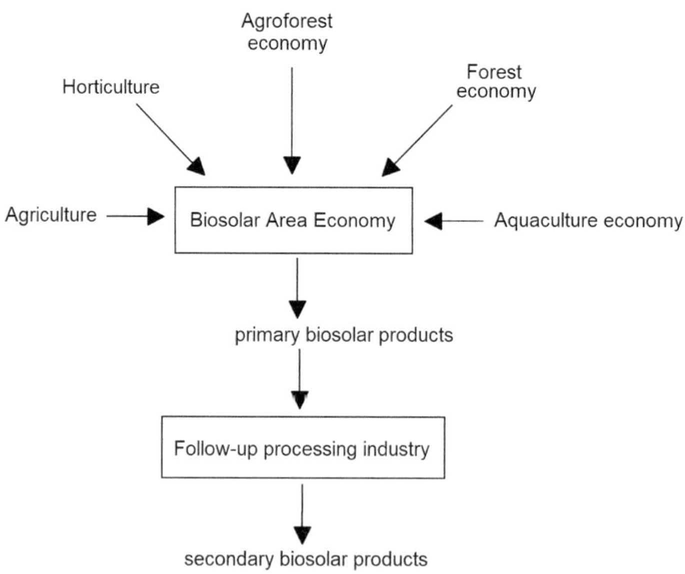

Figure 4.4 Area-related technology fields as a production basis in Biosolar Economy (biosolar area-related economy).

[4]A. Young, 1990; M.S. Smith and C. Mbow, 2014; J.J. Santiago-Freijanes et al., 2021.

4.3 Agriculture as Key Technology for the Realization of Completely Sustainable Closed-Loop Production

While the transition to sustainable energy production in industry is underway, sustainable materials management still plays a minor role in current industrial practice. Despite recycling and circular economy efforts, industrial production largely relies on non-sustainable materials. Consequently, a transformation in materials and resources must follow the energy transition in the foreseeable future. Sustainable industry cannot be achieved solely through the—always inherently incomplete—reuse of materials originally sourced from geological deposits. Instead, the entire energy and materials economy must be fully integrated into natural material and energy cycles, driven by the natural global flow of energy and entropy.

Agriculture holds a key position in addressing this challenge, as its core principle is land management. It is unlikely that the above requirements for linking industry and agriculture can be met solely through the direct interaction between abiotic processes and field crops. In many cases, an industrial application will require a preliminary biotechnological conversion of primary agricultural products. Similarly, many industrial products and the phases of their use will need a subsequent conversion process, particularly one analogous to natural biological reduction processes. Here, biotechnological methods are an obvious choice. In essence, this principle is already being realized in bioenergy production. The challenge for agriculture and industry is to extend this principle to material management. A fully sustainable economy will ultimately consist of three main components: (i) land-based, primarily agrarian-organized primary production; (ii) industrial processing of agricultural primary products; and (iii) the—likely predominantly biotechnological—reduction of industrial products into a form that can be completely reintegrated into land-based cycles.

Clearly, such an economy will require a drastic shift in the raw material base of industry and a significant change in agricultural production. While the shift within plastics production and the production of standard and fine organic chemicals seems relatively feasible, the transformation of currently used mineral resources, especially metals, presents an enormous challenge. Even a highly advanced sustainable economy will likely still need some mineral resources, and certainly metals. However, their use in industry must be minimized as much as possible. The task of land-based production—largely agrarian—will be to provide this necessary minimum of mineral resources and the biogenic organic raw materials. To achieve this goal, there will also need to be a greater expansion of the agrarian principle from land to water surfaces.

A fully sustainable economy uses only products generated through biological primary processes, and all by-products and end-products can be reintegrated into these biological primary processes. Just as the natural biological cycle is sustained through continuous material exchange between autotrophic and heterotrophic organisms—between biological producers, consumers, and decomposers—agriculture must, beyond its traditional role in food supply, fulfill the cycle between industrial consumption and production, including product disposal. In a future sustainable economy, agriculture and industry can be viewed as an analogous pair to the biological complementarity of autotrophic and heterotrophic organisms (Fig. 4.5). Given its land-based production system and the cultivation of photosynthetically active plants, agriculture can be understood as a technical autotrophic system, with industry and consumption forming the corresponding technical heterotrophic system.

4.4 Aqua Farming

Aquaculture is a promising part of the so-called Blue Economy. It has a large potential for sustainable production, which operates under "near-nature" conditions.[5] Aquacultures use sunlight as an energy source. But in many cases, it is not designed to optimize sunlight usage; on the contrary, it frequently relies on intensive, electricity-demanding artificial lighting. Alternative cultivation techniques and land-use strategies—focused on global entropy export—are likely to gain new significance in the context of a sustainable

[5]A.D. Hughes, 2021.

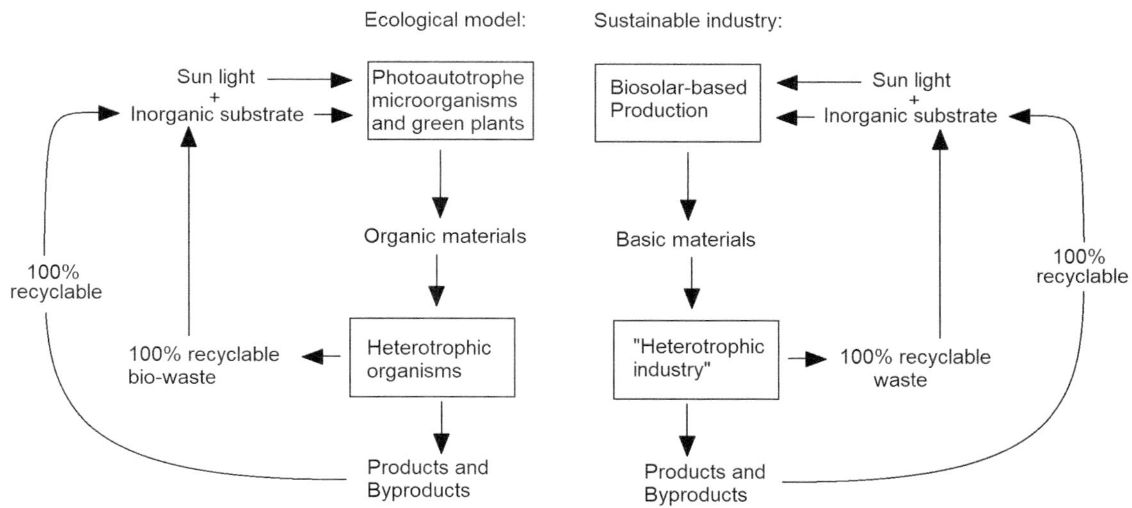

Figure 4.5 "Autotrophy" and "heterotrophy" in an ecological model of biosolar economy and the related material cycles for products and waste.

biosolar economy.[6] These will probably be organized using agrarian principles, representing an extension of agricultural methods onto water surfaces.[7]

The primary motivation for this expansion is the desire to increase the Earth's surface area available for photosynthesis or to expand food production.[8] Additionally, significant will likely be the direct or indirect photosynthesis-driven accumulation of technically needed substances, the enrichment of dissolved or dispersed components in water (Fig. 4.6).

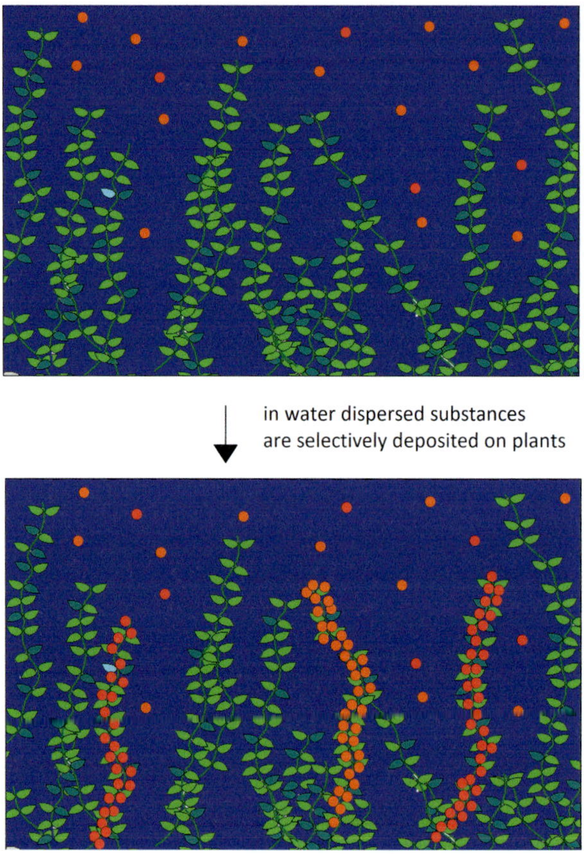

Figure 4.6 Aquafarming-supported selective collection and deposition of substances dispersed in water.

---

[6]J.A. Verreth et al., 2023.

[7]T.V.R. Pillay, 2013.

[8]S. Corrigan et al., 2023.

Most importantly, the concentration of rare substances dispersed throughout the oceans and other bodies of water via aqua-farms is expected to become a central tool for sustainably sourcing these materials. Aquatic cultivation areas powered by sunlight could be designed to accumulate certain ions, molecular materials, as well as dispersed micro- and nanoparticles. Each form of enrichment also acts as a reserve for entropy export, as transforming any dispersed component into a collection or concentrating and sorting it reduces entropy.

As far as plants cannot be applied directly for material accumulation, it seems possible to use microorganisms such as bacteria, which are associated with suited plants, for this purpose. It is imaginable that the efficiency of growth and selective accumulation of special substances can be steered by genetically engineered microbes or plants in the future.

4.5 Metal-Aqua-Farming

The following section will discuss the use of metals as an example of a potential future approach to handling limited resources. This concept envisions raw material extraction, including technically necessary metals, closely integrated into natural material cycles.

In a future sustainably organized industry, the use of metals will be quantitatively reduced compared to current practices. This anticipated decrease in metal usage stems from the fact that most biosystems are incompatible with large amounts of metals or high concentrations of metal ions. Alkali metals like sodium and potassium, alkaline earth metals like magnesium and calcium, and the transition metal iron will likely be exceptions. What differentiates these metals from many others? These five metals are relatively abundant in ionic or bound forms in organisms and readily available in the Earth's crust. These characteristics indicate a relatively high level of biological and ecological compatibility compared to other metals.

However, a modern industry will not be able to entirely do without other metals and semiconductors. Besides their mechanical properties, their unique electron-conducting, redox, and catalytic properties already make them indispensable for complex, highly specialized technologies, and this is likely to remain the case in the future. Nonetheless, their bulk usage can be minimized. For instance, nature demonstrates that the specific properties of metals can be used efficiently at the atomic or ionic level when embedded in an appropriate molecular environment, as in metalloenzymes. While the electronic and catalytic properties of metals and semiconductors, especially their gradation across groups and periods of the periodic table, are challenging to replace with organic materials, other metal functions and properties currently used in technology can often be substituted with organic materials or with biocompatible oxide-mineral components that fit into natural material cycles. This includes properties such as tensile strength and elasticity, temperature, and reaction resistance.

Since mining for ores is fundamentally unsustainable, it will need to be completely phased out in the long term if full sustainability in material management is achieved. Over time, seawater is likely to become the only viable source for sustainable metal extraction, as metals mobilized from land and technical processes through corrosion and leaching eventually end up there, in the form of particles, compounds, or ions.

Seawater contains all elements, including all required metals, though mostly at extremely low concentrations. That means, their distribution in sea water represents a state of maximum entropy. Extraction and accumulation mean a strong reduction of this entropy and demand for high efforts. Extracting metals from seawater on an industrial scale is thus highly energy-intensive and, from an economic point of view, impractical with current methods. However, it is conceivable to shift metal extraction to a biological enrichment process from seawater in analogy to the above-mentioned aquafarming (Fig. 4.7). Such a process would involve communities of organisms that, as components of the biosolar economy, derive their primary energy from photosynthesis. These communities would need to include species that process large volumes of water while selectively binding and accumulating metal ions. Certain organisms, particularly various microorganisms, are capable of metal accumulation and storage, for example, through diverse molecular storage systems known as siderophores.

It may be possible to cultivate species and communities specialized in accumulating specific metals or groups of metals.[9] Genetic engineering could likely enhance the efficiency of metal storage for elements present in only very low concentrations, potentially meeting industrial demand over time.

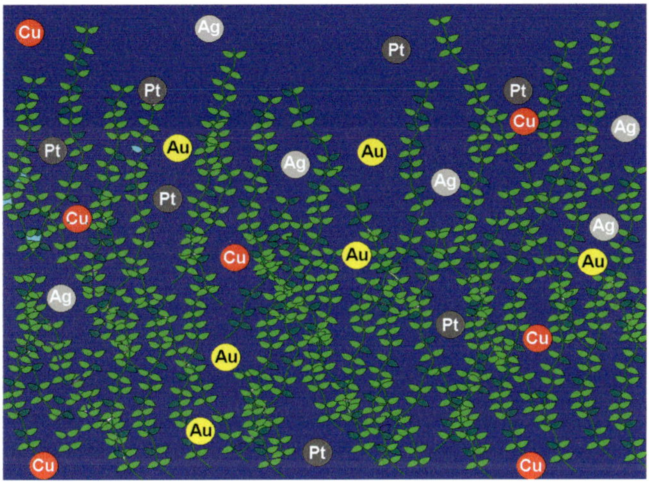

Figure 4.7 "Metal aquafarming": Accumulation of rare metal ions in aquacultures.

One possible approach would be to cultivate aquatic plants in oceanic culture parks, specifically bred for the enrichment of certain metals. These plants would grow similarly to crops, using photosynthesis to increase biomass and producing siderophores that bind metal ions from seawater, even at low concentrations. The necessary entropy export for concentrating the metals would thus be provided using solar energy. Once the plants have matured and accumulated sufficient metal content, they will be harvested. The metals would be separated from the biomass and processed chemically or biochemically according to demand. The organic biomass from these harvested plants would then serve as a byproduct of metal extraction and could be utilized for other material or energy purposes.

[9]Accumulation of heavy metals is evident in aquafarming recently, but represents, firstly, a contamination problem (M.B. Hossain et al., 2023).

An alternative could be biotechnical aquafarming. In this approach, photosynthetic microorganisms would replace green plants. Eukaryotic and prokaryotic microorganisms are suitable for this purpose, including cyanobacteria or unicellular green algae. These could be cultivated in sunlit fermenters arranged as shallow basins or tubes. Another option could involve embedding such fermentation directly into natural environments, using shallow areas of oceans or other standing bodies of water where the organisms grow and can be harvested directly from the water.

4.6 Ecological Embedding of Agriculture

Living systems always depend on energy consumption and the export of entropy. They require dynamics and turnover and perish when cut off from the flow of matter and energy. Sustainability, therefore, does not mean freezing states but rather maintaining vitality, dynamism, and evolutionary potential. This applies not only to crops but especially to the ecological environment into which these crops are integrated. Agriculture inherently involves the substitution of natural communities with crops, creating enforced communities. However, all agricultural land use remains part of an ecotope.[10] Certain ecotopes can be regarded as biocenotic steady states.[11] Ecotopes are thermodynamically open systems and are able to export entropy (Fig. 4.8).

However, sustainable agriculture does not necessarily preserve the original character and composition of the local biocenosis; rather aims to maintain or enhance the fundamental vitality, adaptability, regenerative capacity, and evolutionary potential of the biotope. Although monocultures currently dominate agricultural practices worldwide, future sustainable agriculture could contribute to biotope diversity. Strategies for this may include mixed cropping or the management of small-scale mosaic structures.

---

[10]R.H. Whittaker et al., 1973.

[11]Adaptation of anthropogenically used areas to natural steady-state conditions can protect and promote biological diversity (M.L. Rosenzweig, 2003).

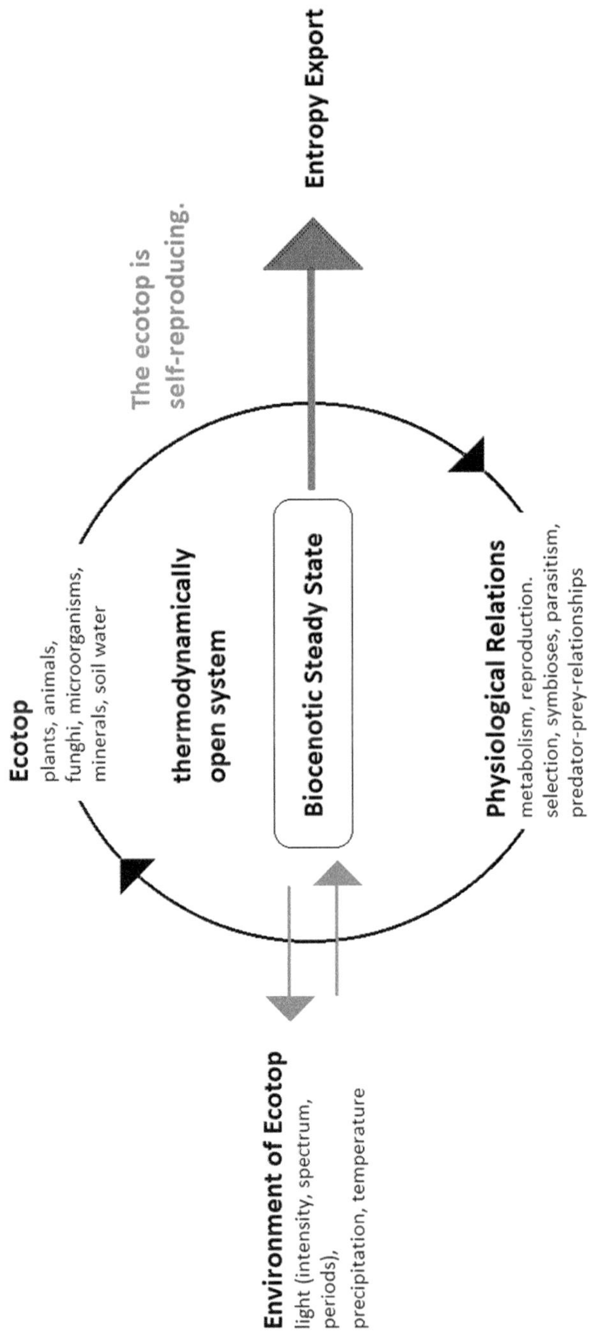

Figure 4.8 Ecotopes as an open system representing a "quasi-biocenotic steady state."

Sustainable ecological integration does not mean returning to some earlier state with its previous communities. That would be an entirely non-ecological perspective. Instead, sustainable ecological integration of agricultural lands means maintaining productivity, vitality, and developmental potential within a biologically and agriculturally productive and adaptive interplay of beneficial and complementary biological components. In this sense, sustainable land use means adapting agricultural practices to the natural components of an ecotope while promoting the adaptation of these natural components to the introduced agricultural elements and the intended production goals.

Agricultural lands are cultural ecotopes. Their management is sustainable only if these cultural ecotopes remain vital and productive over the long term. Generally, the exclusive cultivation of individual species or small groups is insufficient for this. Nor are agricultural ecotopes healthy in the agricultural sense if the composition of communities is allowed to evolve entirely uncontrolled. The inclusion of large portions of the Earth's surface in agricultural management extends the concept of cultural ecotopes to the globe as a whole. The use of the Earth's surface for future food and material production transforms the sum of local cultural ecotopes and regional cultural landscapes into a global cultural landscape mosaic, and thereby, the Earth's natural ecosphere into an agriculture-dominated cultural ecosphere.[12] To sustainably shape this cultural ecosphere, we depend on a high diversity of organisms and ecosystems. Only systems with high diversity—meaning a high level of functional information—can adapt to environmental changes. Therefore, the protection of species, regional and local populations, and the nurturing of highly diverse communities are not contradictory to sustainable use but are prerequisites for the enduring use of our planet's surface. Even purposefully constructed communities for production can exhibit high diversity and play vital roles within the broader ecosystem. Efficient feedback loops are required to develop and monitor the changing biological diversity of different areas and the interplay between near-nature and far-nature ecosystems.[13]

[12]Landscape and ecology: A. Farina, 1998.

[13]D.S. Schimel et al., 2013.

4.7 Co-Evolution in Techno-Ecological Systems

Life is not static; it is in a constant state of change. Sustainability, therefore, does not mean merely preserving a state once achieved but must also include the capacity for further development. The cultivation of certain varieties in large and widespread monocultures contradicts this principle. It has been driven by the desire to achieve maximum production at minimal cost. Nonetheless, even this development of mechanized agriculture shaped by agrochemicals is part of an evolutionary process. However, this evolution occurs primarily at the technical level: machines, methods, and materials are continually refined; responses to disruptive elements in the agricultural biocenosis, such as pests, fungal and viral infections, are constantly improved, and crop varieties and cultivation methods are adapted to changing climatic conditions. The capacity for this technical evolution stems from technological advancements, as well as social and market conditions, and is therefore heavily influenced by factors external to agriculture and relatively independent of local ecological conditions.

Contrasting this technical-cultural evolutionary capability is the actual adaptive capacity of the biological system alongside cultivated plants, namely, weeds, birds, insects, other animals, the mineral substrate of the soil, and the entire edaphon (the complex community of microorganisms and small organisms associated with macro-organisms and soil). Together, these—the technically driven agricultural development and the biological system—form a factual unit that adapts mutually and to other environmental conditions, possessing evolutionary potential. Agriculture inevitably involves the co-evolution of cultural-technical and natural components within a shared ecological system. Sustainable agriculture must be measured by the productivity of this system and its evolutionary potential.

A high evolutionary potential is generally associated with larger, diversified gene pools and species-rich communities of organisms. This creates only an apparent contradiction with agricultural land use, which tends to limit itself to a few, largely genetically uniform varieties. Focusing solely on cultivated plants in a given area overlooks that biological diversity on managed lands is sometimes greater at the microbial level than in natural communities. Contributions to diversity also come from mobile natural components of the environment, the communities in fields surrounding the cultivated area, and other soil organisms that contribute to an area's evolutionary potential. Sustainable agriculture must consider these interactions and the natural resources of the land and its surroundings to accurately assess and sustainably design the ecological value and natural evolutionary potential of cultivated land within the context of technical potential and its evolution, fostering a co-evolution of technical and natural components. This coevolution includes the use of principles of living nature in technologies ("Bionik" systems) as well as the adaptation and genetic manipulation of wildtype individuals and populations for the demands of ecology and production (Fig. 4.9).

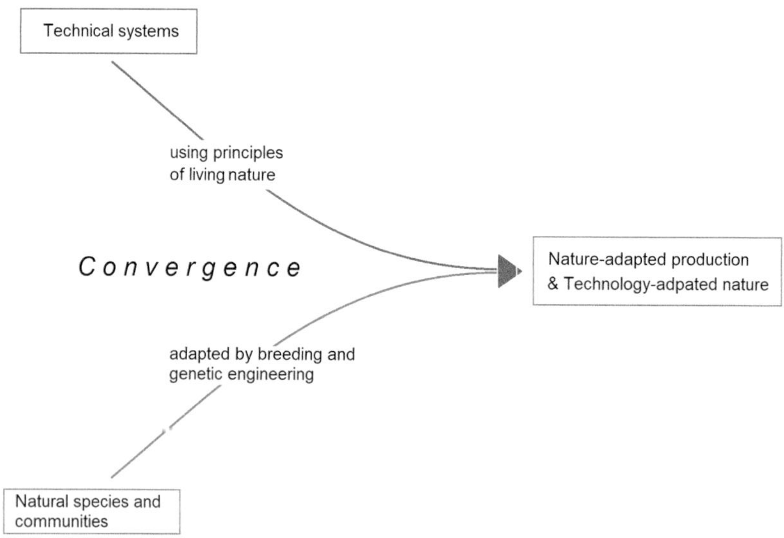

Figure 4.9 Convergence and coevolution of technical and biological systems in nature-adapted, technically used environments.

Sustainable agriculture needs to unite the proven principle of high productivity through varietal purity with strategies for maintaining genetic and biocenotic diversity.[14] This task involves changes in organism communities and diversity within individual natural populations, as well as the development of agricultural methods that incorporate co-cultivation. Short-term agricultural productivity and long-term preservation of biological productivity also require an evolution in cultivation principles. Growing different crops together on small plots often complicates management but significantly increases flexibility in farming practices.

Co-cultivation does not necessarily mean a complete mixing of individual plants. A direct mix of individual plants represents only the lowest level of scale management in co-cultivation, a sort of "individual-dispersed state." The term co-cultivation applies to all mixed-cropping practices that occur on length scales smaller than the typical parcel divisions used in organized agriculture. For example, in large-scale farming over areas measured in kilometers, an ecologically motivated mix of crops at the hectare scale is already a form of co-cultivation. Whether co-cultivation in an individual-dispersed state, clustering of a few plants, or subdividing fields into cultural units on the square meter, square decameter, or hectare scale is optimal depends on site conditions, available management techniques, and agricultural goals.

Mixed-cropping is not a new principle. It is widely used in small-scale farming practices and is a feature of virtually all horticulture. Taking a closer look at gardens and yards in villages and suburbs reveals locally cultivated plant communities—often mixed cultures of edible and ornamental plants—that rarely have identical compositions or quantities, instead presenting a rich, varied picture of diverse "co-cultures." These collectively create a vibrant, colorful, and diverse biocenotic landscape that is human-made rather than developing spontaneously.

Co-cultivation need not be limited to two species or varieties but can involve dozens or even hundreds of types, provided there are ecological reasons and suitable management methods. This approach shows a convergence between agriculture and forestry,

[14]A. Gigon, 1999.

the latter undergoing a shift from productive monocultures to more robust and ecologically valuable mixed cultures that are better adapted and more evolutionarily resilient.

There is only an apparent contradiction between efficient mechanized fieldwork and attention to the meter- and decimeter-scale. Modern measurement and control techniques already allow for capturing each plant in a cultivated area theoretically and treating it individually as needed. Satellite-guided control is practically feasible with precision down to the lower centimeter or even millimeter range. This enables machine-based management of agricultural areas with sub-meter level partitioning, making it no longer necessary to rely on labor-intensive traditional practices or even more demanding horticulture. In the future, modern agriculture equipped with digital technology, sensors, image-based object recognition, and satellite guidance could manage large fields with complex crop structures in an automated manner.

These capabilities significantly expand the scope for field partitioning, co-cultivation, and complex mixed-cropping systems. This also opens up entirely new options for cropping strategies that closely resemble natural ecosystems in species composition, are more resilient to plant diseases and drought, and are richer ecosystems overall.

For future sustainable agriculture, it is essential to understand the role that crop mixtures and field partitioning scales play in the efficiency of agricultural use. Basic research should investigate how optimizing land partitioning and crop mixtures can connect diversity, productivity, and sustainability. This research should also explore how agricultural techniques in finely partitioned fields can be supported by computer-aided machine guidance, using, for example, multi-sensor systems, image processing techniques, real-time object identification, individual plant recognition, and high-precision GPS guidance.[15]

[15]Digital tools play an increasing role in realizing an efficient agriculture and are very promising for the improvement of sustainability in agricultural production; for example, review of Apps (S. Inwood and V.H. Dale, 2019).

4.8 Reproduction and Storage of Information in Entropy Management

The entirety of living organisms in an area, along with the genetic information that is potentially available, constitutes an ecologically invaluable reservoir of negentropy. The information stored in the gene pool of the local community of organisms has been accumulated through biological entropy export during the preceding development of this community. It is the product of changing environmental conditions and varying inter-organismic relationships. This pool of information enables the community at a given site to benefit, for example, in the event of environmental changes. This informational wealth is contributed by not only vegetative, i.e., currently physiologically active organisms, but also by dormant life forms (Fig. 4.10).

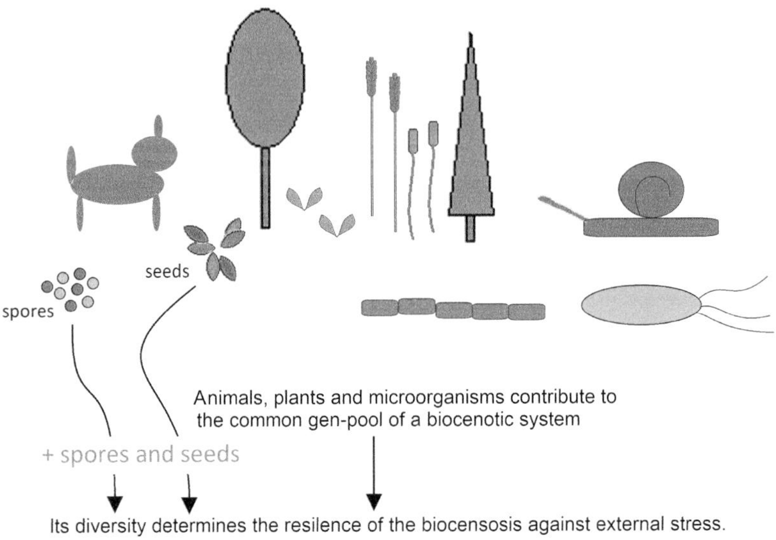

Figure 4.10 Active organisms as well as dormant forms as spores and seeds contribute to the genpool of an ecological system and its resilience.

Even agricultural systems with few species and varieties can still hold ecological value if suited to the local conditions. The Earth's surface abounds with species-poor biocenoses. Expanding productive agriculture into vast areas with rainfall deficits, poor soil formation, specific nutrient-poor soil conditions, and challenging climate is a crucial task. Solving this challenge is significant not only for a sustainable economy largely driven by agriculture, but also makes the ecosphere richer, more productive, ecologically diverse, more evolutionarily capable, and resilient. Current efforts in "desert agriculture" must also be assessed according to sustainability criteria and biological productivity, specifically, their ability to support soil formation, water retention, and efficient conversion of solar energy into biomass.

Many sites with natural communities that appear species-poor at a particular time are often defined by a relatively small number of vegetative forms. However, they frequently exhibit a high number and diversity of dormant life forms. This includes long-lived seeds, fungal and bacterial spores, and other dormant stages. The ability to develop a wide range of vegetative forms that can activate from dormant stages under certain and sometimes sporadic environmental conditions is also the result of evolutionary adaptation. This principle is rarely applied in agriculture but could become a valuable tool for making less productive areas of the land surface usable within biosolar land management.

In such land management, the principle of scale-diversified co-cultivation would incorporate dormant life forms in a hierarchical, site-specific layout. This would involve a distribution of different vegetative and dormant life forms, adapting to the respective local conditions.

4.9 Dynamic Sustainability

The original concept of sustainability refers to an economic approach in which resources are utilized and maintained in such a way that consistent use is possible even in the distant future. This is often associated with the idea of a steady-state, where what is used naturally regenerates, allowing the current state to be projected into the future. This corresponds to the model of a system that constantly reproduces itself despite being utilized.

A modern concept of sustainability must include the capacity for the regeneration of economically utilized natural systems, but must also consider the evolutionary arrow of time inherent in all natural systems, including ecosystems. It is not only the ability to produce economically viable yields that must be preserved but also the capacity to change, adapt to altered environmental conditions, and develop new characteristics.

It is entirely evident that the areas used for production in agriculture, forestry, and aquaculture are far from being habitats where natural systems are sustainably reproduced. On these lands, production can occur over extended periods through mechanical cultivation, sowing, fertilization, and harvesting, but sustainability in the sense of maintaining and developing ecological and evolutionary potential cannot be achieved in these areas. Sustainable agriculture requires convincing protocols for the flow of energy and materials, integrated landscape planning and management respecting economic as well as ecological conditions, a reliable monitoring, and suited integration of agricultural activities into the social environment.[16]

It is, therefore, an illusion to achieve sustainability in the long term within the framework of intensive land use. Either the ecosystems on certain areas develop sustainably—meaning that humans extract components from natural systems only to the extent that their natural regeneration, adaptation, and development potential is genuinely preserved—or land is utilized for efficient production, but in doing so, natural ecosystems are replaced by artificial systems without adaptive and developmental potential.

To find a way out of this dilemma, two complementary strategies likely need to be pursued to achieve dynamic sustainability.[17] On the one hand, land use must be able to alternate between phases of intensive utilization for effective production and phases of natural—predominantly spontaneous, ecological—development. On the other hand, production methods must be developed that increasingly bring artificially created and utilized ecosystems closer to natural systems with their ecological potential.

---

[16]V.H. Dale et al., 2013.

[17]L.B.L. Amui et al., 2017.

This would enable humans in the future to adapt synthetic ecosystems increasingly to the properties and dynamics of robust and simultaneously productive natural ecosystems. Conversely, humanity will also claim the freedom to adapt natural ecosystems more and more to its needs without diminishing their adaptive and evolutionary potential. Both strategies together represent a convergent development of natural and synthetic ecosystems. This idea will be explored further in the next chapter.

4.10 Area Competition and Development of Area Structure: Fluctuating Mosaic Patterns

The management of time scales will be crucial for a fully sustainable global economy. This timescale management encompasses not only entropy management, involving material and energy economies, but also the management of the Earth's surface and land use. Here, too, transitions must be allowed and encouraged to ensure the flexibility required for long-term viability. The currently dominant, permanently regulated division of land by states into protected areas and zones designated for exploitation contradicts this need.

The landscape resulting from future agriculture will be shaped primarily by three key characteristics: (i) hierarchical land segmentation, (ii) site-specific ecological embedding, and (iii) scale-diversified co-cultivation. For the landscape, this implies a revival of the mosaic principle, which has proven effective in many historical and prehistoric cultures. An ecologically healthy, productive, and resilient landscape will consist of a well-thought-out mix of natural and cultural elements in varying proportions. It will form a patchwork of individual areas, with a structure influenced both by natural geographical conditions and by technical land-use requirements and history (Fig. 4.11).[18] This mosaic-like land-use pattern means that the characteristics of individual patches will change over time on different timescales. The composition of individual mosaic units and the overall layout

[18] A. Heinimann et al., 2017.

will no longer be determined solely by local economic needs, but will also meet the requirements for securing sustainable regional land management and a globally sustainable economy over the long term. The ecological and social values of mosaic-like structured landscapes are confirmed by studies on land use by traditional agricultural societies.[19]

Figure 4.11 A mosaic-like landscape pattern supports an efficient area use, matching ecological, and economic requirements at the same time.

The landscape elements will vary in terms of the intensity of technical management, the diversity of species communities, the permitted level of spontaneity in population development, cultivation conditions, crop rotations, ecological functions for the surrounding environment, regional ecological potential, and for the ecosphere as a whole. Essential for sustainability and long-term economic viability are the robustness and adaptability of the entire mosaic structure. At one extreme, individual landscape elements may take on the character of nature reserves or strict

[19]For example, area use and landscape segregation in Laos (J.C. Castella et al., 2013) and in Cote d'Ivoire (N.J.K. Kassi and G. Decocq, 2008).

conservation areas. At the other extreme, monoculture plots are not excluded. In between, there are areas with varying levels of species mixtures, ranging from intensively managed non-cultivated zones to moderately used and managed biotopes with largely spontaneous population development. Thus, the individual areas within the mosaic differ significantly in terms of the intensity of use and the degree of technical and cultural influence. Vitality, productivity, and sustainability are ensured by the ecological interplay in the optimal composition of the mosaic and the dynamic potential within the individual sections, as well as by the modification and evolution of the mosaic structure itself.[20]

4.11 Flexible Protection of Environment ("Schutz- und Nutz-Wechsel")

The envisioned picture of the future, highly segmented, ecologically rich mosaic landscapes, stands in strong contrast to the current legal division of the Earth's surface into areas where nature is protected and areas that are largely available for arbitrary use, where aspects of biodiversity and ecological resilience are subordinated to economic interests. In extreme, there are areas completely removed from human use, where nature is left entirely to itself on one side; on the other hand, areas are defined where unlimited exploitation is allowed (Fig. 4.12). This division might cater to bureaucratic convenience in spatial organization, but it scarcely meets the ecological requirements of a healthy and dynamic environment in the future.

The interaction between shifting environmental conditions and the transformation of species communities is fundamental to Earth's ecological dynamics. The stress caused by short-term fluctuations in external conditions and the resulting ecosystem responses through gradual adaptation promote genetic diversity, enrich species communities, and stabilize the ecosphere as a whole. Therefore, a permanent division of land into constantly cultivated, highly productive areas where only a few crops rotate, on one side, and nature reserves that are expected to remain

[20]Concept of mosaic pattern: J. Schmithüsen, 1948.

unchanged, on the other side, is not only unnatural but, in the long run, also ecologically inappropriate.

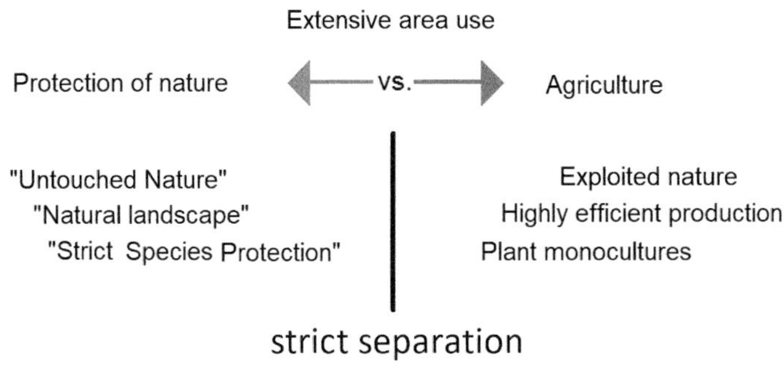

Figure 4.12 Recent land management and nature conservation are frequently based on a strict and long-term separation of free-usable areas, on the one hand, and strictly and long-term protected areas, on the other hand.

Although it is ecologically sensible to preserve total reserves for the protection of species and biocoenoses, many ecologically significant areas, such as numerous forests, can simultaneously serve practical purposes. Thus, function-switching between protected and productive areas should not be taboo. It seems better to develop concepts including changes between protection and use on different time scales ("Schutz-und Nutz-Wechsel")[21] (Fig. 4.13).

However, these transitions must be managed so that each regional landscape structure, with the sum of its component areas, remains productive while also maintaining high biological diversity. To achieve this, such transitions must especially consider the timescales of natural processes within evolving ecosystems. Pioneer stages, succession, and maturation often require centuries or even millennia, so the alternation of conservation and use must sometimes also be planned on these longer timescales. This suggests that agricultural strategies need to align more closely with forestry principles, including the principle of forestry sustainability—applied, however, to the entire mosaic of areas—as well as the planning timelines and species compositions.

[21]German; meaning: change between protection and economical use of areas.

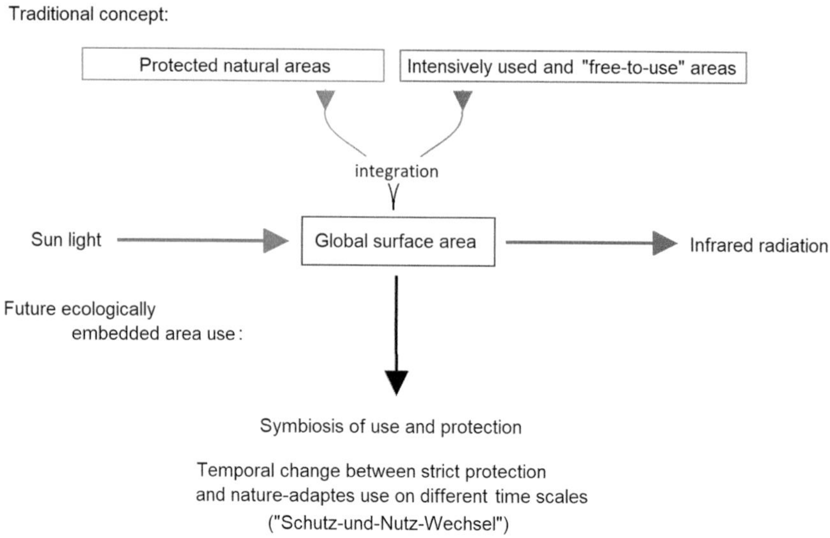

Figure 4.13 Integration of strategies in natural protection: integration of strict protection and free-use of areas by a model with temporally changing strictness of protection.

Trees and other long-lived environmental components are essential to many ecosystems, and the interrelationship of agricultural standard crops with the accompanying organisms of agricultural ecosystems, trees, and their associated communities will play a central role in future sustainable agricultural landscapes. For instance, park-like landscapes with cultivated and grazed sections, as seen in historical cultural landscapes, might prove to be advantageous.

4.12 Mineral Potential, Soil Nutrient Balance, and Nutrient Accumulation

Integrating industrial production and waste management with natural material and energy cycles should preferably occur through interfaces suited to land use patterns. Only in this way can a sustainable export of entropy from technical and cultural systems be maintained over time. Air, water, and soil are the three central connecting elements. Therefore, technical raw materials, by-products, and end products must be considered together with the nutrients needed for biological growth. In a fully sustainable economy, all these components, including every nutrient required on agricultural land, must be linked to natural material cycles. Industrial entropy production, besides consuming high-grade energy and releasing heat, also results in material dissipation. In sustainable resource management, only those substances produced and released into the environment by industry should be ones that can be fully absorbed by living nature and accumulated, transformed, and utilized within the biospheric material cycle through natural processes, directly or indirectly powered by solar radiation.

The coupling between industrial production, agricultural processes, and spontaneous natural processes involves numerous chemical and biochemical transformations of materials. To ensure sustainable recycling of all involved material components, quantities, and bonding forms of the participating chemical elements in the technical and natural branches of these cycles must be harmonized.

Carbon and hydrogen are universally available as basic elements with short cycle times through atmospheric CO_2 and soil moisture. Nitrogen deficiency can be compensated for by suitable crop rotations and co-cultures that promote nitrifying soil or root bacteria, enabling nitrogen fixation from the air, thus facilitating nitrogen exchange on short timescales. More problematic are other mineral components such as phosphate, potassium, calcium, magnesium, iron, and certain trace metals, which are plentiful in some soils but lacking in others, and are therefore supplemented today by mineral fertilization based on raw materials coming from mining (Fig. 4.14). These unsustainable sources must be substituted in the future. This demand is challenging because huge amounts of fertilizers are required in agriculture.

There are three main pathways for adding these mineral components to soils: (i) dust transport through the air, (ii) natural transport via animal movement, or (iii) technical extraction from renewable sources or seawater, followed by conventional transport and application as mineral fertilizers. In most cases, a mixed strategy would likely be the most practical. The natural

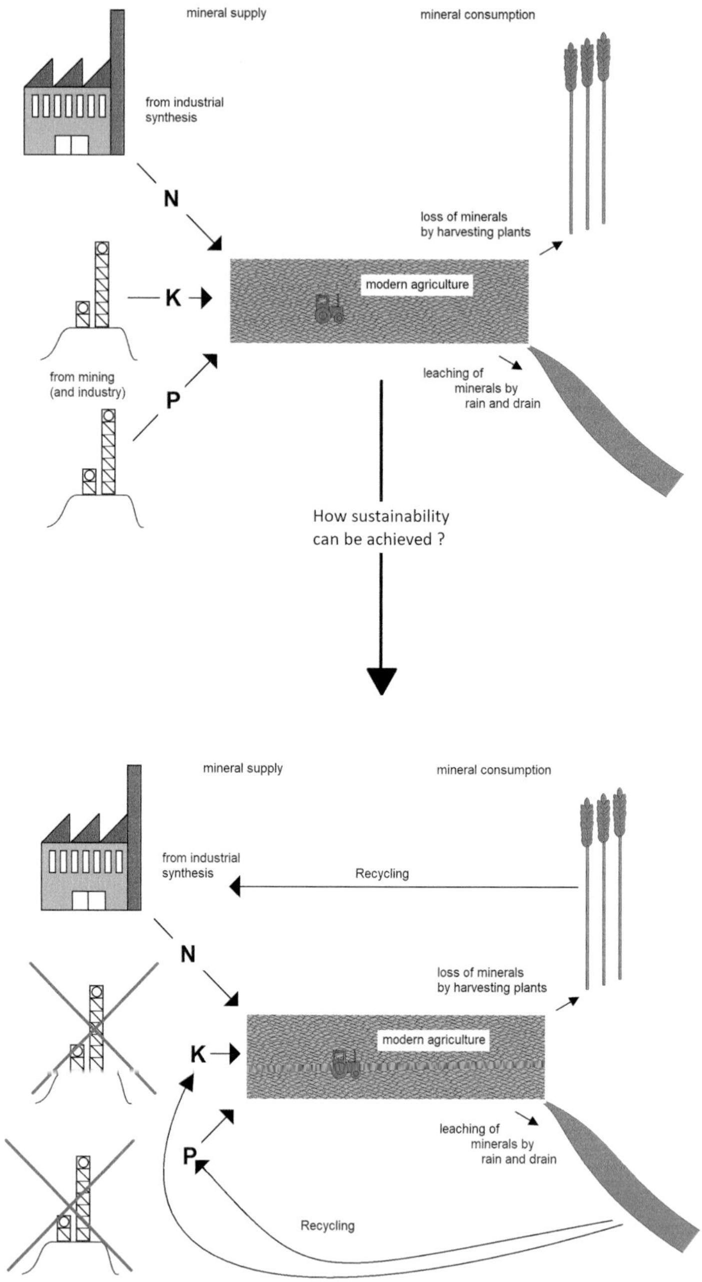

Figure 4.14 Need for substitution of mineral fertilizers from mining by fertilizers from cycling processes.

transport of dust through atmospheric circulation is an important means of mineral input, though it is difficult to control regarding both intensity and direction, as well as the specific components needed. This component may play a role in future global climate management. The architecture of land mosaics could contribute by designating spatially and temporally specific areas with mineral-rich substrates to function as dust sources. Their contribution to mineral supply could be local and regional, for instance, transferring between adjacent land elements with geologically different substrates. An example would be the frequent pairing of acidic, silica-rich sandstone soils and basic, calcium- and magnesium-carbonate-rich limestone soils. Such controlled natural dust fertilization could also extend across larger areas and distances, such as the continental or even transcontinental air transport of mineral-rich desert and steppe dust with high iron content.[22] However, any inclusion of such airborne transport in a mineral supply concept must consider that undesirable components—both inorganic and biological, such as seeds, fungal spores, and bacterial spores—could also be transported, which is difficult to control. Moreover, mass dust mobilization requires open areas, as in desert regions, which ideally should be reduced through improved local biological productivity.

The vision of controlled global mineral transport could, in principle, be extended to volcanic ash, which is often very rich in minerals valuable for fertilization. Since ancient times and continuing today, this resource has been utilized passively, with the land surrounding volcanoes often serving as fertile and highly productive agricultural areas. It is conceivable, however, that humanity could one day develop tools to moderate volcanic activity, controlling the timing, intensity, and duration of eruptions. This anticipated capability would serve not only to improve safety during eruptions and earthquakes but also to support the regional, continental, and global dispersal and deposition of minerals for sustainable agricultural use.

Closer to home and simpler to implement are strategies to prevent excessive loss of essential minerals from agricultural

---

[22]N. Meskhidze et al., 2005; F. Lambert et al., 2015.

land, enhancing natural accumulation and bioavailability. These factors can, in principle, be influenced through appropriate compositions of biocoenoses, providing a strong case for smart crop rotations and mixed cultures. For instance, siderophores in plants and microorganisms, or the biogenic precipitation of mineral components in soils, can contribute to an ecologically balanced mineral supply and thus soil fertility.

Nevertheless, natural erosion, leaching, and product extraction will always result in mineral loss. This can be partially offset by promoting mineral release from the soil substrate, for example, through occasional deep plowing. However, this practice does not represent a sustainable solution. A durable and sustainable concept for the targeted supply of soil with the required huge amounts of mineral components remains an enormous challenge.

4.13 Soil as Natural Basis – Management of Its Resilience and Evolutionary Power

In few other environments is "learning from mistakes" as clearly visible as in the functioning of soils. Soils do not represent individuals or populations in a biological sense but are, from structural, material, and biological perspectives, highly heterogeneous and diverse entities. This makes them distinctive components of a habitat, rich with biological and ecological information accumulated through soil formation processes. The information stored within soil is a product of current and past life processes, which involve the export of entropy. Thus, soils serve as a form of negative entropy, providing ecosystems with developmental potential and creating the foundation for efficient responses to changes in environmental conditions, thereby supporting ecological robustness. The information pool of soil includes both mineral and biological components.

Soil structure and the pool of organisms adapted to the soil form an ecological memory[23] within the soil. The information stored in this memory also embodies a reserve of entropy export.

[23] J.F. Johnston et al., 2016.

This richness is largely contributed to by microorganisms, especially those found in low individual numbers but with high diversity. This memory is represented by the coexistence of vegetative (i.e., metabolically active) organisms, individual cells, and dormant forms.[24] All these are products of previous environmental conditions in which the respective organisms could grow and reproduce. Changes in environmental conditions have always been accompanied by a quantitative redistribution of different types. The formation of resting and dormant forms allows certain types to survive long periods under unfavorable growth and reproduction conditions in every soil. The complementarity of relatively few active types that occur in large quantities at a given time and many types that appear in very small numbers and are either minimally active or inactive represents the soil's experience base, preserves "learned" information, and forms a kind of systemic informational set analogous to an organism. Destroying this information means a loss of structure and response capability, equivalent to an increase in entropy.

The presence of a growth-promoting mineral mixture and the availability of essential elements in soils are just one aspect of soil conservation and care. Even within the inorganic portion of the soil, numerous properties are crucial for general fertility, as well as for the development of specific plants and life communities. For instance, grain structure, porosity, water transport behavior, swelling capacity, adsorption/desorption properties of mineral component interfaces, and erosion behavior all play a role alongside the chemical composition.

Added to this is the entire highly complex biological component of the soil, including free organic compounds, biogenic macromolecules, plant roots, fungi, multicellular microscopic organisms, yeasts, bacteria, and protists. Besides other organisms, humus soil can contain up to about one billion bacterial cells in one gram of soil material. Besides a few highly abundant bacteria, a high number of less abundant species are present in soil, typically (Fig. 4.15). The interaction of mineral components, water, and these diverse organisms with the above-ground parts of

[24]A. Canarini et al., 2021.

plants and the animals living there shapes ecological conditions. Soil is an incredibly complex substrate that constitutes the entire life function of a location and thus fundamentally determines its biological productivity and resilience.

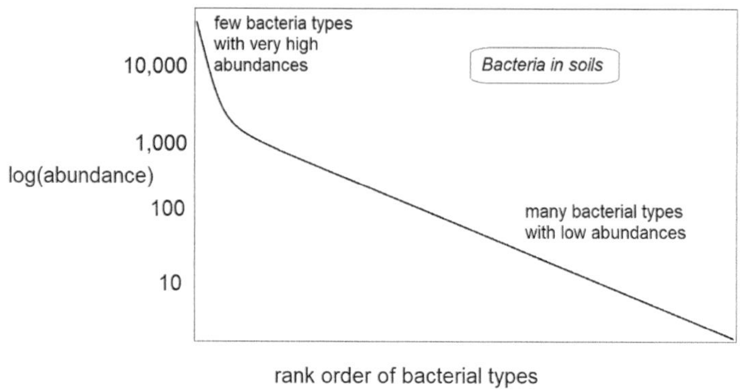

Figure 4.15 Typical abundance distribution of bacteria in soil (principal rank order diagram, arbitrary numbers).

The above-ground visible diversity of plant communities is only a small part of the overall diversity of life communities, to which soil organisms make a substantial contribution. This is also why soils play an important role in sustainable agriculture. Their metabolic performance is part of the total biological output. Their adaptability is crucial for the overall adaptability and evolutionary potential of a site. Soil care, therefore, cannot be limited to mineral and water management. Soil structure and especially soil organisms are among the most critical resources for agriculture. The interaction of plant roots with fungi, small organisms, and bacteria is what defines the vitality of soil.

The complexity of the organism composition in most soils across different locations corresponds to an enormous variability in soil composition. The composition of these life communities holds a large amount of information and a high degree of order. The loss of this diversity means an increase in entropy. The biogeochemical potential of the totality of soils on the Earth is a tremendous ecological treasure. At the same time, each soil, especially its upper layer, is a highly dynamic entity, with structure, chemical, and biological composition constantly changing in response to environmental conditions and seasons. The ecological information stored in soil communities is a product of negentropy accumulation during the evolution of each biotope and its entropy export activity in the past. Preserving this diversity and dynamism and making them useful for the broad range of tasks in future agriculture as part of a sustainable economy is essential.

It seems practically impossible to archive and conserve the diversity and development potential of today's soils. Freezing the current state of soils is neither feasible nor purposeful. Therefore, future land use by agriculture must establish frameworks that do not necessarily preserve the state of individual soils but rather maintain the immense diversity of soils on the Earth and allow for and promote their continued development.

Land use must consider the preservation and development of microbial resources and promote adaptability within microbial communities. This includes the demand that land use be designed to support the evolutionary processes of soil microorganisms and their communities. Studies on soil bacterial metagenomes show that biological productivity and diversity do not always correlate. Often, disturbed natural communities show a higher diversity of soil bacteria than well-developed natural ecosystems. Agriculture must thus allow for the fluctuation of natural conditions that stress ecosystems and, at the same time, can promote evolutionary processes, particularly for the microorganisms involved, through a suitable level of ecological stress from soil tillage and varying uses. This approach can thereby increase diversity in individual life communities and globally.

For this purpose, a sub-area structure and the architecture of agricultural crops, along with variably expressed life communities, are particularly suitable. This general area structure concept combines the allowance of spontaneous soil development in natural and near-natural areas with a form of soil development adapted for efficient agricultural production, which, however, must be managed so that even for intensively cultivated sub-areas, the sustainability of soil use remains ensured.

4.14 Hierarchical Character of Structured Area Use

Natural landscapes often lack clear structuring. Where there are no pronounced geomorphological boundaries, such as coastlines, mountain slopes, or varying geological substrates, that create distinct changes in local environmental conditions, transitions within biotopes tend to occur gradually. This often results in large, contiguous areas of relatively homogeneous vegetation. The forests and tundras of northern Asia serve as a vivid example of this. The spatial distribution of trees and specific communities seems to be more or less random, reflecting the absence of a local and regional organizing principle from outside. At first sight, natural landscapes contain mostly no specific information for a surface subdivision beyond natural geographical factors.

However, this view is not completely right. Even untouched ecosystems are marked by a hierarchical organization of ecological relations and spatial distributions. They are only less visible than in cultivated landscapes.[25]

In contrast to untouched nature, cultural landscapes look entirely different. Agriculture and other land use factors force a clear structure in the area use. Production conditions, transportation, trade, and social structures, in addition to natural spatial factors, shape the layout. The division of agricultural land parcels represents only the lowest level of a hierarchy of surface sizes, which is layered further by settlement structures. The landscape is divided into areas of affiliation and regions preferred for specific types of agricultural, forestry, or even commercial use. In many cultural landscapes, a settlement size hierarchy is reflected in the land structure, ranging from single farms to hamlets, villages, small towns, mid-sized centers, large cities, and metropolises.[26] The hierarchies in predominantly agricultural landscapes are rooted in long-standing traditions.[27] Historical examples of elements larger than individual parcels include

[25]F. Klijn and H.A. UdoDeHaes, 1994.

[26]W. Haber, 1994.

[27]G.D. Peterson, 2002.

village communal fields, clearing zones around individual villages or groups of villages, settlement chambers, forest areas reserved for lordly hunting, or even extensive, outwardly bounded urban fields. As a result, even in historical cultural landscapes, a pronounced hierarchy of land divisions emerged, forming not only a pattern created based on economic and legal considerations but also a mosaic of cultural biotopes. For as early as the Neolithic period — about 4,000 to 8,000 years ago in Central Europe — a mosaic pattern in agriculturally used areas is already presumed. The hierarchically structured land embodies a local and regional organizing principle and its associated information. Embedded natural local communities, such as hedgerows or small woodlands, and agricultural land structured for farming purposes embody structural information and represent a state of reduced entropy compared to a random landscape with a stochastic distribution of communities. This structural information provided a functional advantage in early agricultural use compared to random land layouts. It was economically, socially, and ecologically beneficial.[28]

With the mechanization of agriculture and the associated consolidation of land parcels, this complex land structuring was largely abandoned for economic reasons, and many structuring landscape elements were removed. While this allowed for more efficient agricultural production, it also reduced the ecological and, often, the societal value of a landscape, for example, for tourism. The loss of structure in historical cultural landscapes is a loss of systemic information. Compared to the finely structured historical cultural landscape, the largely undivided, large-scale agricultural landscape represents a state of higher entropy.

The hierarchical structure of ecological systems—natural as well as agricultural systems—is continued downward from the smallest agricultural area spots to the macro- and microstructure of soil and the single macro- and microorganisms. Opposite the two-dimensional hierarchy in land elements, there exists a hierarchy in three-dimensional soil structures that ranges from

[28]Recent traditional agricultural practices can also include valuable elements, which might be usable for the development of sustainable agriculture in the future (J.C. Ruiz and C.M. Yanez, 2014).

the scale of large rock fragments and macropores (centimeter to millimeter range) through the size of minerals and micropores (sub-millimeter to micrometer range) down to nanostructures, on the scale of nanoparticles, molecules, and atoms.

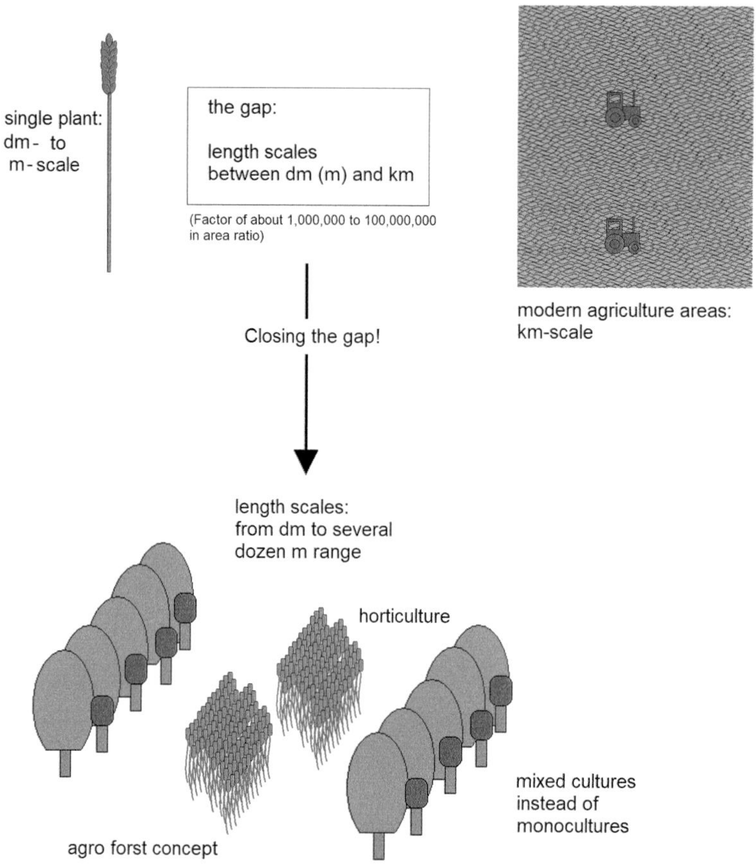

Figure 4.16 New technological concepts could help to close the size gap between single plants and typical areas in modern agriculture by managing an efficient land cultivation in mosaic structures.

Currently, a structural gap exists along the length scale between soil structures (around cm to dm) and the plot sizes in modern agriculture (around 0.1–1 km). This gap, absent in some traditional agricultural uses, commercial horticulture and in residential gardens, may gain renewed attention in the future organization of sustainable agriculture (Fig. 4.16). This section of the length scale—covering about four orders of magnitude in length and consequently eight orders of magnitude in area—could see renewed focus. Ecological land use, the integration of production, recreation, and conservation, as well as increased attention to landscape aesthetics, are likely to lead once again to greater structuring and thus an accumulation of structural information, i.e., a reduction in entropy within the landscape.

Chapter 5

Transformation and Continuity

5.1 Will to Global Acting

The global dimensions of land use, the exploitation of oceans, and the industrial consumption of natural resources have become a worldwide transgression. The age-old human trust that our interventions in nature would, sooner or later, be offset or at least tolerated without significant changes to natural living conditions has been rightly shattered. The drastic violation of sustainability principles, through the mass exploitation of fossil fuels and raw materials, large-scale conversion of natural landscapes into cultural ones, and widespread release of byproducts like CO_2 into the atmosphere, oceans, and soil, has created a state that overwhelms natural regulatory mechanisms. The current state, reliant on continued resource consumption, cannot be sustained indefinitely.

Would a radical solution of forgoing technology help? Can the wheel of progress be turned back? What would a drastic reduction in industrial entropy production mean? A return to the pre-industrial economy and lifestyle would inevitably push the vast majority of people into conditions of low material standards

Entropy and Sustainability: Challenges of Future Energy and Material Management from a Thermodynamic Point of View
Michael Köhler

ISBN 978-981-5129-99-1 (Hardcover), 978-1-003-74889-2 (eBook)
www.jennystanford.com

DOI: 10.1201/9781003748892-6

of living. Mechanical labor would have to be replaced by manual work, and animal labor would regain importance. A much larger portion of the population would need to engage in food production, drastically reducing the manufacturing of consumer goods. Gross domestic product and living standards would plummet. Agricultural yields would likely fall significantly with reduced use of fertilizers and pesticides. Meat consumption would be severely limited to ensure adequate nutrition through plant-based foods. Cultural, educational, and scientific spaces would be restricted, leading to a decline in educational standards, deteriorating hygiene, and compromised medical care. Life expectancy would decrease, and child mortality would rise. Supply shortages and the breakdown of urban social structures would create severe social tensions. In sum, drastic reductions in industrial production would lead to a marked decrease in quality of life, an increase in individual life risks, and high societal risks. This is something the vast majority of people do not want.

Entropy production due to high energy and material turnover is an absolutely necessary consequence of modern industrial society. To maintain functional social structures, we must accept the entropy production associated with production, transport, and communication. Therefore, the primary task of industrial society is not to suppress entropy production altogether, but to organize it sustainably. This should consist of two components: (i) optimizing currently used entropy-generating processes, and (ii) seeking new, fully sustainable strategies.

The measures for the energy transition follow precisely this demand. On one side, energy savings without fundamentally limiting entropy-producing processes, and on the other side, expanding sustainable energy sources by substituting fossil fuels with sources linked directly to the Sun's energy flow through the Earth to the cosmic background.

At the heart of optimizing existing processes is the principle of efficiency: saving does not mean renouncing certain services, but minimizing the resources required. This principle applies not only to energy consumption but to the reduction of overall entropy production. The efficiency principle is a fundamental property of systems far from equilibrium. Even simple abiotic systems follow developmental pathways that move from areas of high entropy production toward reduced entropy production. The key parameter here is entropy production density, the entropy generated per unit of time and volume. According to Glansdorff and Prigogine, all non-equilibrium systems tend toward a state of minimal entropy production density. Non-equilibrium steady states, which are always associated with entropy production, generally stabilize through minimizing entropy production.[1] A physical example is the synchronization of coupled oscillating systems with different natural frequencies: phase differences create potentials, whose equalization leads to energy loss and entropy release. These processes bring the phases closer together, and with sufficient coupling strength, synchronization is achieved. Once synchronized, phase differences and sums of elongations over time intervals are minimal. A relatively small energy flow, and thus low entropy production, is sufficient to maintain the synchronized state. Thus, after synchronization, a state of minimal entropy production is achieved.

Reducing production costs, saving energy, and reducing material use are typical expressions of reducing entropy production density. Today, this includes more efficient machines, better-insulated buildings, higher efficiency in energy production and conversion, low-power computers and communication systems, fuel-efficient vehicles and aircraft, and improved logistics. It also includes deposit systems, recycling materials, and reducing the release and spread of pollutants and products in the environment. Such entropy production minimization finds broad consensus, as it benefits producers and consumers. The general insight into the need for cost savings through more efficient operations and the general drive for efficiency support this positive attitude. Thus, minimization of entropy production density within existing structures is generally desired.

Much more critical, however, is finding social consensus for structural changes that reduce entropy production.[2] Are there ways to bridge radical changes in entropy production with current industrial strategies? Rational discussion and sound decisions

[1] I. Prigogine and P. Glansdorff, 1971.

[2] T. Jackson, 2018.

on the right path are only possible if there are clear standards for assessing risks and opportunities. Clear ideas are needed about how technological developments unfold, what fundamental structural changes in technology and society are expected, and what risks exist.

The concept of structure accumulation, i.e., of negative entropy (negentropy) within the involved systems, can help search for evaluation standards. This concept involves preserving, maintaining, and caring for all existing structures necessary for functionality while allowing for the development and integration of new structures. Only this way, the proven mechanisms have been retained, new mechanisms developed, and flawed structures gradually replaced. This applies to all functional structures such as biological structures like habitats and communities, settlement structures from single houses to megacities, industrial structures from small workshops to large-scale chemical plants, economic structures from small businesses to international corporations and trade networks, and social structures from families to internet communities, essentially, all the varied structures at global, local, and regional scales.

The necessary shift to a fully sustainable economy must follow a semi-conservative principle: ecologically and socially necessary functionalities must be preserved. The structures and mechanisms that underpin them must be protected until they can be replaced by something more suitable. Implementing this principle requires critical analysis, ethical evaluation, broad discourse, and systematic action.

Decisions must be based on value standards that protect current unsustainable practices if they are functionally necessary, and force a sufficiently rapid transition to procedures that improve sustainability and provide a path toward complete sustainability. The simultaneous necessity of protecting entrenched, unsustainable practices and the need to overcome them poses a very serious problem. Solving it will require difficult compromises and decisions that cannot always be made with the value standards currently dominant in our society. The problem is compounded by the diversity of values and interests among the world's different human societies and within groups within these societies.

This dilemma calls for the primacy of a comprehensive functional ethics. On the one hand, this ethics encompasses the consistent preservation of established structures if they are not responsible for functional problems that could be resolved by new, less problematic structures. On the other hand, this functional ethics includes creating spaces for new functional structures and substituting outdated ones.

Despite the difficulties, developing ethical consensus remains a solvable problem. Discussions about such a consensus will always reflect historically inherited social structures and their value systems, which, despite internal contradictions, hold significant functional importance almost everywhere. However, these discrepancies can be overcome without abandoning valuable functional structures if ethical principles are prioritized based on functionality. From a Western perspective, these are mainly built on three principles: (i) the freedom and dignity of the individual, (ii) the freedom of economy and trade, and (iii) the organization of international relations under historically established national and international law.

These three principles undoubtedly have legitimacy and functional importance within developed industrial societies. However, they largely overlook the fundamental importance of the complexity of functional hierarchies and the accumulated information in functioning systems, reducing these to the individual, state, and market levels. This shortfall is most apparent in the lack of ethical integration of the living environment and natural resources. Despite the "green thinking" of recent years, the preservation of the long-term capacity for entropy export, the protection of global and local ecosystem functions is essential for sustainable material and energy cycles, and the associated local and regional social structures seem often to be not clear understood and secondary to individual freedom, the market, and powerful states.

Regarding the necessary technical and societal restructuring, this represents a distortion in the value standards we need to make specific decisions. The will to act must be accompanied by a correction of this ethical imbalance.

The long-term goal of a completely sustainable economy requires as its highest maxim the organized development of the necessary functional structures: social, economic, and ecological. The ethical standards underlying today's human, economic, and international law must be complemented by ethical standards for social, economic, and ecological structures, prioritized in service of long-term sustainability. With an understanding of the importance of hierarchical structures and the systemic information inherent in them for sustainable functionality, an ethical reassessment of these structures and their ranking on the value scale is necessary.

From a sustainability perspective, for example, it is simplistic to define nature conservation primarily through species protection. The presence or absence of a species in a habitat may indicate the habitat's ecological condition, but in a dynamic ecological system where habitats shift and communities continuously change, this is only of limited significance. More important than protecting local populations and species is the diversity of varieties and breeds, the diversity of communities, the diversity of dormant life forms, openness to local and regional environmental changes, the activation of dormant forms, migration, dispersal, and replacement of active communities. We need ecosystem robustness as well as plasticity, adaptability, and developmental potential, the potential to adapt to changing conditions.

A high capacity for adaptation to future, unpredictable changes, and maintaining development potential despite change is crucial for biological systems and their ecological function on the Earth. The same applies to the needed dynamism and adaptability of human society. We need individuals open to change and willing to engage in the necessary restructuring of structures. But beyond that, we also need functional social groups in their diverse forms, from friendships and life partnerships to families, clans, village communities, tribes, and nations, to work teams, employee associations, interest and professional groups, political parties, lobbyists, and internet community platforms.

All structures of nature and human culture on the Earth are not merely products of random events. Rather, they are the result of accumulated information, a long-term shaping and development of functionality that, despite the highly stochastic course of history, generally follows a selection based on success, functionality, adaptability, and optimal coupling to other structures. In this sense, these hierarchically ordered structures embody functionally accumulated negentropy, too. They embody the arrow of time, the irreversibility of natural processes, and the Earth system's capacity to gather experience, learn, and build complexity. In this way, they also represent the complex capability of life on the Earth to organize entropy export with increasing efficiency and to leverage this entropy export in turn to generate ever more complex structures, creating a broader array of variability and potential for information. The transition from our highly unsustainable industrial society to a future society that operates sustainably on a global scale must be integrated into this process of learning and accumulation.

It is entirely clear that the very different social structures that have developed on the Earth cannot be simplified into a uniform code of laws, nor a single, linear value scale. Reasonable ethical standards must therefore differentiate by current situation and future requirements, by capability and available resources, by immediate need and sustainability, and by historical and traditional, cultural, and religious conditions. In these criteria for shaping a nuanced scale of values, the existing functional structures and their expected importance for future development are reflected (Fig. 5.1).

Even though not everything can be precisely predicted, it is nonetheless clear that, independent of historically evolved structures and value systems, the criterion of sustainable action, ensuring the survival and ongoing development of humanity within the global ecosystem, must be at the top of this hierarchy of values. Beneath it, other values are arranged with consideration for established structures and regarding their varying potentials for necessary transformation and future functionality. The diversity of ethical standards and value systems across human societies does not primarily present a problem to be resolved through standardization. Rather, this diversity should be viewed as an opportunity, a reservoir of experiences and structures accumulated in the evolution of diverse human societies. However, these structures and associated mechanisms must be subject to a functionally oriented ethical assessment. A global will to act presupposes a comprehensive ethics of structures.

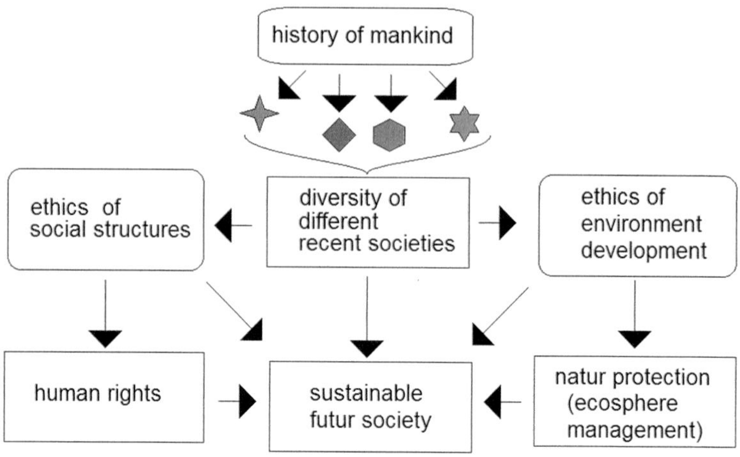

Figure 5.1 Insights into ecological requirements must be mirrored in different social structures, experiences, and ethical standards on the Earth to achieve a global sustainable action.

5.2 Ability to Acting

Humanity is currently in a phase where it is beginning to develop a collective will for global action, aimed at enabling the long-term sustainable development of humanity on the Earth within a robust global ecosystem. This represents an important step toward a shared will for change. However, the will to act must be accompanied by the capability to act to translate intentions into action. To achieve this, two sets of prerequisites must be met. First, the technical capabilities for action are required; second, sufficient room for action must be available.

These two prerequisites can be divided into areas of responsibility. Science and technology are tasked with creating the technical prerequisites for action, expanding the range and effectiveness of technical possibilities, improving the prediction of the consequences of decisions and actions, and continuously enhancing the criteria for assessing the purpose and appropriateness of technical measures. Meanwhile, economics and politics are responsible for creating the space for action, facilitating decision-making, enforcing these decisions, providing the necessary resources, and ensuring that the consequences of these decisions are bearable for society.

Room for action is not created by renouncing technology and advancements in natural sciences but through the development of new technical solutions and deeper insights into the structures and laws of nature. Scientific and technical progress is an integral part of cultural development. It reflects the conversion of a portion of our entropy export potential into the accumulation of information.

We need technological progress. A renunciation of technological development is justified only when it involves significant and uncontrollable risks. Any other form of restriction of science and technology by humanity equates to a renunciation of information, structure, and complexity, as well as developmental and adaptive potential, competitiveness, and survival capability. Such a path would go against the arrow of time, lead to regression, and is therefore inherently doomed to fail. The transition to complete sustainability needs a consequent transformation process that is open to new technologies and promotes innovation and economic growth. It demands the development of a "knowledge-based" bioeconomy.[3]

However, an increase in freedom to act cannot be achieved by disregarding existing, functional structures, even within a contemporary framework, nor by revaluing all values or breaking with all traditions. Radical cuts invariably result in the abandonment of valuable existing tools, methods, and information, thereby reducing available options and potential for action. While unsustainable technologies must be replaced, ceasing production entirely and starting anew is counterproductive if essential supply chains are disrupted, technological discontinuities arise that cannot be compensated for, or economic and social structures collapse. Radical cuts may create short-term opportunities but are often accompanied by structural and informational losses that

---

[3]A. Pyka and K. Prettner, 2018.

outweigh the gains. Therefore, a longer transition phase must be expected, in which some traditional technologies will be switched off, others will be modified or substituted by new technologies. In addition, new non-complete sustainable technologies will be introduced, if they are necessary, as bridging technologies. All these technical changes are accompanied by the use and further development of technical and social experiences of the different societies on the Earth (Fig. 5.2).

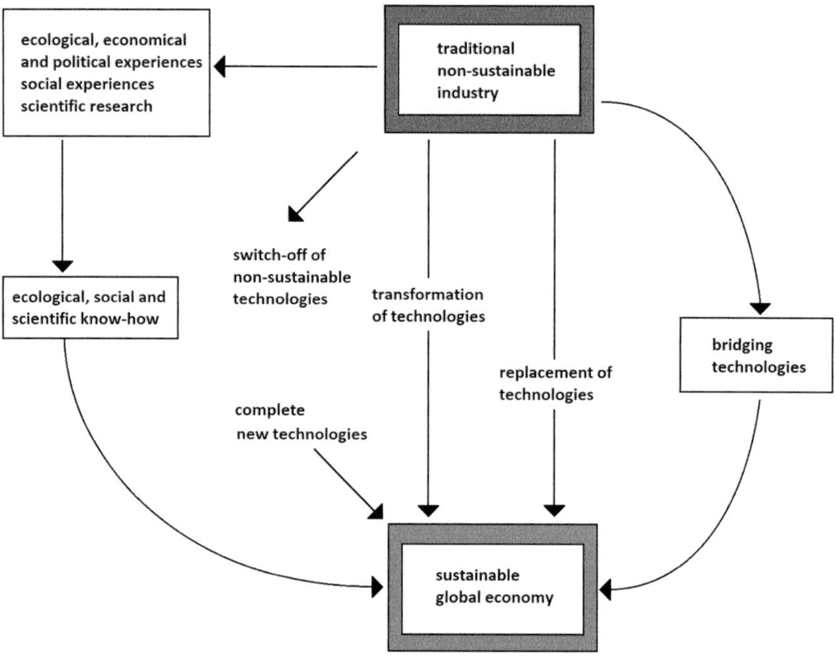

Figure 5.2 Combination of strategies in the transition between the non-sustainable industry and the future sustainable global economy.

Technical capability for action emerges through the combination of scientific and technical progress, the parallel use of existing technologies, and the development of transitional technologies. These transitional technologies ensure the maintenance of vital existing structures and processes when outdated technologies need to be replaced, but new technologies are not yet available.

The complementarity of technological tradition and progress in enhancing technical capabilities is paralleled by the complementarity of social and economic progress with established social and economic structures. Recognizing the value of social structures at all hierarchical levels and understanding the consequences of social changes are crucial prerequisites for achieving the economic and social capabilities necessary to shape a sustainable global economy, for coming to a worldwide sustainable living "Green Society" (Fig. 5.3).

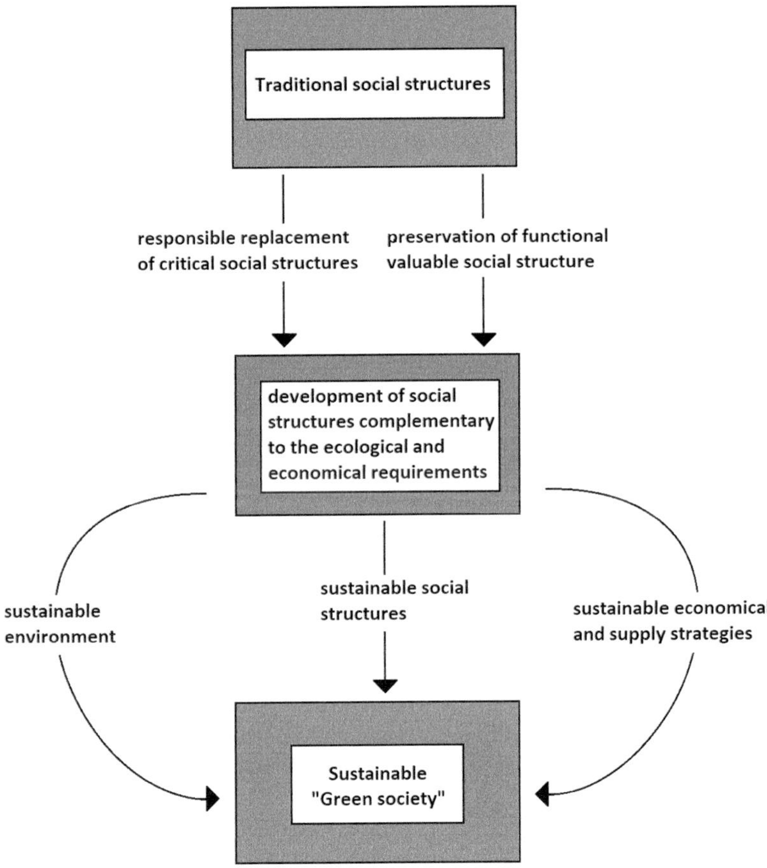

Figure 5.3 From traditional social structures to a future "Green Society": a challenge to achieve sustainability for a healthy environment, stable social structures, and powerful resource development.

New social orders and economic models can enhance societal capacity for action. To achieve this, existing economic systems must be adapted, and outdated social structures must be replaced. However, radical restructuring can be dangerous, fraught with significant risks, and may even lead to catastrophic conditions. Every dismantling of structures initially results in a loss of order, in a sense, an increase in entropy within the system. Yet, the capacity to act requires order and structure. Therefore, calls for social change must be grounded in respect for the functionality of structures that have arisen from local or regional conditions. Only by preserving and enhancing this functionality and integrating it into the formation of new global structures that serve the goal of ecological global governance can the ability to act on a global scale be achieved. For this reason, it is essential to complement the ethics of human rights with an ethics of social structures in the interest of sustainable development. Consideration of sociocultural factors must become an essential part of the development of an efficient bioeconomy.

5.3 Provision for Future Generations: Construction of Entropy Export Reserves

The resource-intensive nature of our current industrial society corresponds to the image of humanity as the primary disruptor of the global ecological balance. This impression is justified and compels humanity to act in the interest of a healthy biosphere and its survival. However, this perception should not obscure that humans are not the sole factor contributing to the destabilization of the global ecosystem. The idea of long-term stable ecosystems, enduring natural equilibrium, constant environmental conditions, and a harmonious interplay of plants, animals, fungi, and microorganisms on the Earth is an illusion.

Even without human interference, the environment changes, and seemingly stable and harmonious living conditions are disrupted. Most striking are sudden geological or astronomical catastrophes, perceived as natural disturbances such as earthquakes, volcanic eruptions, chaotic changes in solar activity, and the impacts of celestial bodies on the Earth.

Yet, stochastically occurring events are far from the only natural causes of disruption to a constant and harmonious environment. Convection currents in the Earth's mantle and the slow but continuous drift of continental plates, the precession of the Earth's axis, and the interplay of these processes with the Earth's annual orbit, gradual changes in ocean currents, and rapid shifts in atmospheric circulation and tides combine to create a mixture of periodic and episodic events on different timescales. As we know from the Earth's geological and climate history, these changes can lead to global ecological transformations, triggering dramatic upheavals—ecological catastrophes.

Added to this is the dynamic nature of living systems: the evolution of populations and species, habitats, and communities. The timescales of these processes range from the generation times of fast-replicating microorganisms (about 20 minutes) to millions of years needed to form many new genera and species, which alter the biological composition and ecological character of entire habitats. The living world itself, and its coupling with abiotic processes, harbors a vast potential for dynamism, transformative power, and sometimes unpredictable instability.

All these changes, such as cosmic, geological, and biological, are expressions of the irreversible dynamics of nature. While the movement of celestial bodies, the daily course of the Sun, and the annual alternation of seasons, vegetation phases, and growth cycles suggest recurrence, periodicity, and uniformity, these regular changes are merely a modulation superimposed on the permanent, directional change that affects all systems and processes. The seemingly unchanging periodicities are only the transient external appearance of the relentless arrow of time, which governs all processes that do not represent a true equilibrium. All real change is ultimately irreversible. Every event or process that recurs in our environment is fundamentally based on an irreversible dynamic. Behind this lay the continual dissipation of thermodynamic potential and the production of entropy. Only where systems can export entropy to their environment and compensate for or even overcompensate their entropy production can they learn from the past, build structures, accumulate information, and store resources to adapt to future changes.[4]

A key feature of organisms and ecosystems is their ability to respond to environmental changes through adaptation. The biological systems on the Earth have accumulated and stored knowledge about such changes over their long evolutionary history. The diversity of short- and long-term environmental changes has led to an accumulation of explicit information in genetic datasets and implicit information in the interactions between organisms in a habitat and their relationships with soil and atmosphere. Biological systems have learned to respond to changes on different timescales, although their responses are constrained by the inherent timescales of their dynamics. The resulting changes in biological systems under external stress are more dramatic the more severe the environmental changes.

One important outcome of evolutionary learning processes is the precautionary behavior of biological systems. Examples include adaptations to regular, shorter-term changes, such as the seasonal shedding of fur, storing food for winter, seasonal growth patterns, and mating behaviors. Other examples of evolutionary learning have longer-term effects and are not limited to individual organisms, species, or populations. These include soil formation and water regulation through leaf litter, as well as the development of dormant life forms, such as seeds and spores, which not only enhance the long-term survival and distribution of individual species but also ensure the viability and resilience of entire biotic communities under changing environmental conditions.

These results of biological evolutionary learning also represent a form of structure-building and the accumulation of negative entropy (negentropy). Ultimately, they integrate into the larger global geoecological processes that collectively ensure the relative long-term robustness of the global environment.

At first glance, it may seem illogical in the context of evolutionary thinking that organisms with lifespans of only a few decades, or at most centuries, contribute to optimizing processes that unfold over thousands, millions, or hundreds of millions of

---

[4] I. Prigogine, 1955.

years. And indeed, teleological thinking, which attributes the formation of coal deposits and the creation of oil, gas, and other reservoirs to some inherent purpose of supporting human development and industrial society, is completely disconnected from reality. Nevertheless, the long-term, large-scale sequestration of atmospheric carbon in the form of biogenic carbonates and fossil fuels has significantly contributed to shaping the global climate and maintaining moderate average atmospheric temperatures as we experience them today. Such long-term behavior cannot be explained as the outcome of a single evolutionary event but requires the concept of long-term collective learning. This, in turn, typically involves multiple phases of changing biological system characteristics and varying environmental and selective pressures.

Indeed, the Earth's history has been marked by repeated cycles of glacial and interglacial periods and phases of high and low greenhouse gas concentrations. These shifts in selective pressures for populations and communities have partially driven the development of broadly carbon-fixing ecosystems. The carbon sequestration associated with the formation of these reservoirs represents a form of structural organization and global, biologically induced order. These reservoirs, not only fossil fuel deposits but also various mineral deposits, are the result of the biological activities of countless organisms throughout the Earth's history, reflecting their ability to export entropy and generate order. These structured deposits are now industrially exploitable, allowing us to benefit from their reduced-entropy state and harness the associated potential for technical processes. Hence, we utilize these biogenic reservoirs as reserves of biologically generated entropy exports.

Transitioning to a fully sustainable global economy means not only largely refraining from the use of these reserves accumulated over geological timescales but also regenerating such resources and creating new entropy export reserves for potential future challenges. If humanity sees itself as an organic part of a learning biosphere, we must also learn to use today's and tomorrow's potential for long-term precautionary measures.

Currently, as far as we can foresee, much of the entropy flow from the Earth to space does not contribute to building structures and order on the Earth but is lost. While the biosphere harnesses some of the energy-rich solar radiation for photosynthesis, a significant portion is converted to heat on land surfaces and in the upper layers of the oceans. Another substantial share is simply reflected into space, primarily by clouds and ice. Only a minuscule fraction of the energy continuously available on the Earth is utilized by humanity through photovoltaic systems, wind power, and photosynthetically generated biomass.

In this context, the global netto entropy export $dS_{\text{out, earth}}$ is primarily consumed by total entropy production S_{prod} on the Earth. Additionally, we can still rely on some of the entropy export reserves deposited throughout the Earth's history $S_{\text{expoldreserve}}$ to satisfy our entropy export demand. This results in the following current entropy flow balance:

$$dS_{\text{out, earth}}/dt + dS_{\text{expldreserve}}/dt = dS_{\text{prod}}/dt. \quad (5.1)$$

Short-term sustainability requires that we do not consume more entropy export reserves than can be newly created. Exploitation of reserves must cease entirely:

$$dS_{\text{expldreserve}}/dt = 0. \quad (5.2)$$

If some exploitation of old entropy export reserves remains unavoidable, it must be offset by the creation of new entropy export reserves $S_{\text{expnewreserve}}$ for keeping the entropy flow in balance:

$$d|S_{\text{expoldreserve}}|/dt = d\,|S_{\text{expnewreserve}}|/dt. \quad (5.3)$$

This leads to the following overall balance for the available entropy export potential:

$$d|S_{\text{out, earth}}|/dt + d|S_{\text{expoldreserve}}|/dt - d|S_{\text{expnewreserve}}/dt = d|S_{\text{prod}}|/dt. \quad (5.4)$$

Long-term sustainability, however, requires precautionary measures and the creation of additional entropy export reserves. This lowers the principal potential for a sustainable entropy production:

$$d|S_{\text{out, earth}}|/dt + d|S_{\text{expoldreserve}}|/dt - d|S_{\text{expnewreserve}}|/dt > d|S_{\text{prod}}|/dt. \quad (5.5)$$

Besides the continuous improvement to exploit the principle possible potential of entropy production S_{prod}, we must ensure the development of efficient strategies to build new entropy export reserves. What might such entropy export reserves look like? In principle, there are no significant limits to imagination at this stage. We cannot yet predict what types of entropy export reserves will be needed and useful. However, it is reasonable to assume that humanity's ability to utilize a wide variety of entropy export reserves will grow with increasing technical capabilities. The generation of entropy export reserves by the recently living generations should at least balance the ecological burdens from the previous generation and the recently generated ecological burdens (Fig. 5.4).

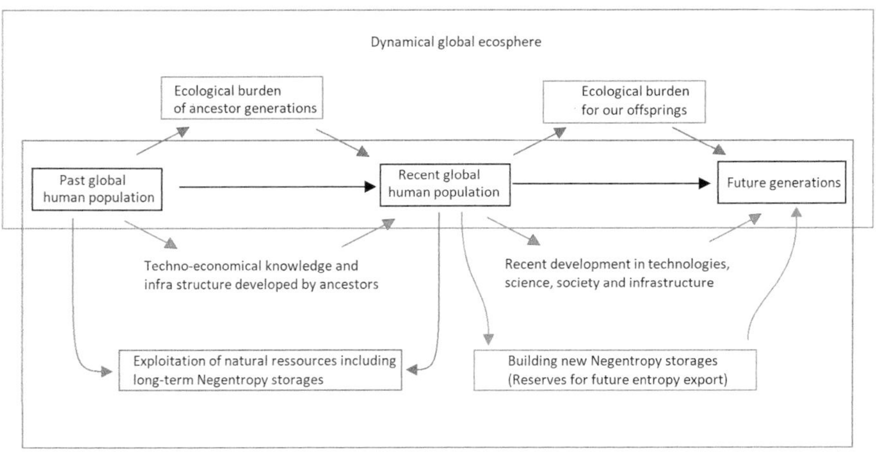

Figure 5.4 Ecological burdens and needs for creation of new entropy export reserves for future generations.

From today's perspective, two approaches seem particularly relevant: (i) the long-term sequestration of carbon from the atmosphere or technical processes in concentrated forms and (ii) the accumulation of materials that could be advantageously used industrially, as fertilizers, or in other distributed applications.

Examples of entropy export reserves already exist, even if their creation has not been motivated by sustainability concerns. The extraction of deposits containing valuable raw materials in low concentrations and the subsequent concentration of specific components represent a process of negentropy accumulation. The extraction and concentrated storage of gold, for instance, serves as an example, driven by its use as a currency equivalent.

5.4 Transition Phase for Achieving Global Sustainability

Just as the mechanisms of life processes impose temporal limits on the ability of organisms and ecosystems to learn and adapt, so too are technical and organizational adjustment processes for transitioning to sustainable economic practices subject to minimum time requirements. Fundamentally, technical systems can be modified much more quickly than ecological developments or biological evolution occurs. However, recent history in technology and economics shows that the development and maturation of technologies require a certain amount of time. For instance, the modern automobile has a development history of over a hundred years, and even modern solid-state electronics and integrated circuit technology have taken about six decades to reach their current level. Chemical plants typically have lifespans of several decades, and their foundational developments often require a comparable time frame when considering their predevelopment phases.

Therefore, the technical steps toward a sustainable global economy will not be completed within the timeframe of a legislative period or a five-year plan. Rather, these are tasks spanning several generations. At first glance, this might appear to ease the urgency of the required changes. Unfortunately, the opposite is true. No time can be wasted in initiating the next steps. Precisely because we cannot overhaul the entire economy and technology in one grand effort, and because key technologies are still completely lacking, and because we need transitional technologies that will serve as stopgaps for a long time, the situation is pressing.

In the coming decades, during which these changes must take place, global ecological problems will continue to worsen, and the burden that we, like our parents and grandparents, the generations of global industrialization, are placing on future generations will grow significantly. During this period, we must not only create the technological prerequisites for sustainable economic practices but also address the already evident negative consequences of ecological mismanagement by previous generations. Moreover, we must prepare pathways for future generations navigating this technological transition, enabling them to cope with the cumulative ecological and technological burdens.

5.5 Living with Permanent Changes: Sustainability Without Constancy

The concept of "entropy," introduced by Rudolf Clausius, quantitatively describes the fact that in a spontaneous process, energy always flows from a warmer body to a cooler one, but never the other way around. The reason for this lies in the movements of the smallest building blocks of matter, atoms, and molecules. The random transfer of countless mechanical impulses in the microcosm is the cause of the thermodynamic directionality observed in the macroscopic world. This arrow of time determines the direction in which physical and chemical processes unfold. It dictates that we live in a world of irreversible change.

The irreversibility of processes in the real world is not limited to the microcosm of atoms and molecules. The processes of living nature demonstrate how fundamentally the arrow of time shapes our world. Plants develop from seeds, grow, bear fruit, age, and die. The life cycle of animals and humans follows the same principle. The path always leads from conception, through embryonic development, youth, adulthood, maturity, and aging, to death, but never in the opposite direction.

Even the seemingly eternal images of the stars, the endlessly recurring movements of the Sun and planets, cannot escape the effects of the arrow of time. If we observe cosmic processes on a sufficiently large time scale, the illusion of permanence collapses like a house of cards. Everything that appears to be an immutable constant in everyday life ultimately reveals itself to be transient upon closer inspection. Constellations distort, stars change color, eventually vanish, and new stars are born. The universe expands, our Sun ages, and according to all we know, it will, in a few billion years, first heat the Earth to deadly temperatures and then engulf it in a gigantic fireball.

Yet even on much shorter time scales, our world is in constant flux. Natural ecosystems, like our cultural landscapes, are subject to continuous change. Life and the Earth as a whole are constantly evolving. The arrow of time prevails, no matter how strongly we try to resist it. That is why responsible and sustainable action does not mean preserving an observed or imagined state, but rather ensuring the ability to live and develop amid change, maintaining readiness for what is expected, and openness to the unforeseen that the future will bring.

Our pursuit of security, stability, and the preservation of cherished, pleasant, and seemingly practical conditions fundamentally fails against the relentless force of the arrow of time. We cannot resist the causally successive steps of natural processes. Constant change is the fundamental principle of the world. One can either despair over this fact or embrace it. Uncertain futures and transience are the fate of human existence and the defining characteristic of our world. The German poet Johann Wolfgang von Goethe captured this idea in the final stanza of his poem Blessed Longing:

"… Und so lang du dies nicht hast,
dieses "stirb und werde,"
bist du nur ein trüber Gast
hier auf unsrer Erde.[5]

In the translation of Robert Bly:[6]

"… And so long as you haven't experienced this:
to die and so to grow,
you are only a troubled guest
on the dark earth."

---

[5] "And until you have possessed this 'die and become,' you are only a gloomy guest on this darkened Earth."

[6] https://www.leonardcohenforum.com/viewtopic.php?t=22251; last download February, 14th, 2025.

References

Adám, B.; Goen, T.; Scheper, P.T.J., et al.: From inequitable to sustainable e-waste processing for reduction of impact on human health and the environment, *Environ. Res.* 194 (2021), 110728.

Amui, L.B.L.; Jabbour, C.J.C.; Jabbour, A.B.L.D.; Kannan, D.: Sustainability as a dynamic organizational capability: A systematic review and a future agenda toward a sustainable transition, *J. Cleaner Prod.* 142 (2017), 308–322.

Arrhenius, S.: Über die Dissoziationswärme und den Einfluss der Temperatur auf den Dissoziationsgrad der Elektrolyte, *Z. Physikal. Chem.* 4 (1889), 96.

Arrhenius, S.: Über die Reaktionsgeschwindigkeit bei der Inversion von Rohrzucker durch Säuren, *Z. Physikal. Chem.* 4 (1889), 226.

Arrhenius, S.: Über den Einfluss des atmosphärischen Kohlensäuregehalts auf die Temperatur der Erdoberfläche, *Bihang* 22 (1896), Stockholm.

Atkins, P.: *The Second Law*, New York, 1984.

Birner, R.: Bioeconomy concepts, in: *Bioeconomy* (ed. I. Lewandowski, Springer Open), 17 (2018).

Boltzmann, L.: On the relationship between the second fundamental theorem of the mechanical theory of heat and probability calculations regarding the conditions for thermal equilibrium, *Sitzungberichte der Kaiserlichen Akademie der Wissenschaften. Mathematisch-Naturwissen Classe. Wien, Abt.* II, 76 (1877), 373–435.

Boltzmann, L.: *Populäre Schriften*, Leipzig, 1905.

Brillouin, L.: The negentropie principle of information, *J. Appl. Phys.* 24 (1953), 1152–1163.

Brook, B.W.; Buettel, J.C.; Hong, S.: Constrained scenarios for twenty-first century human population size based on the empirical coupling to economic growth, arXiv:2109.14209.

Canarini, A.; Schmidt, H.; Fuchslueger, L. et al.: Ecological memory of recurrent drought modifies soil processes via changes in soil microbial community, *Nat. Comm.* 12 (2021), 5308.

Carlowitz, H.C. von: Sylvicultura oeconomica. *Anweisung zur wilden Baum-Zucht*, Leipzig, 1713.

Castella, J.C.; Lestrellin, G.; Hett, C.; Bourgoin, J.; Fitriana, Y.R.; Heinimann, A.; Pfund, J.-L.: Effects of landscape segregation on livelihood vulnerability: Moving from extensive shifting cultivation to rotational agriculture and natural forests in northern Laos, *Human Ecology* 41 (2013), 63–76.

Clausius, R.: Über die bewegende Kraft der Wärme und die Gesetze, welche sich daraus für die Wärmelehre selbst ableiten lassen, *Poggendorffs Annalen* 79 (1850), 500–524.

Clausius, R.: Über verschiedene, für die Anwendung bequeme Formen der Hauptgleichungen der mechanischen Wärmetheorie, *Annalen der Physik und Chemie* 125 (1865), 353–400.

Corrigan, S.; Brown, A.R.; Tyler, A.R.; Wilding, C.; Daniels, C.; Ashton, I.G.C.; Smale, D.A.: Development and diversity of epibiont assemblages on cultivated sugar kelp (saccharina latissima) in relation to farming schedules and harvesting techniques, *Life* 11 (2023), 209.

Dale, V.H.; Kline, K.L.; Kaffka, S.R.; Langeveld, J.W.A.: A landscape perspective on sustainability of agricultural systems, *Landscape Ecology* 28 (2013), 1111–1123.

Davis, P.C.W.: *The Physics of Time Asymmetry*, University of California Press, 1977.

Dyksma, S.; Bischof, K.; Fuchs, B.M. et al.: Ubiquitous Gammaproteobacteria dominate dark carbon fixation in coastal sediments, *ISME J.* 10 (2016), 1939–1953.

Ebeling, W.; Feistel, R.: *Physik der Selbstorganisation*, Berlin, 1982.

Einstein, A.: Über die Entwicklung unserer Anschauungen über das Wesen und die Konstitution der Strahlung, *Physikalische Zeitschrift* 10 (1909), 817–825.

Fahr, H.-J.: Mit oder ohne Urknall. *Das ist hier die Frage*, 2. Aufl. Berlin, Heidelberg, 2016.

Farina, A.: *Principles and Methods in Landscape Ecology*, Chapman & Hall, London, New York, 1998.

Gigon, A.: Agricultural sustainability does not imply biocenotic sustainability, *Appl. Veget. Sci.* 2 (1999), 89–94.

Guth, A.H.: *The Inflationary Universe: The Quest for a New Theory of Cosmic Origins*, 1998.

Haber, W.: System ecological concepts for environmental planning, in (F. Klijn, ed.) *Ecosystem Classification for Environmental Management*, Kluwer Academic Publishers, Dordrecht, The Netherlands, 1994, 49–67.

Hawking, S.: *A Brief History of Time*, 1988.

Heinimann, A.; Metz, O.; Frolking, S.; Christensen, A.E.; Hurni, K.; Sedano, F.; Chini, L.P.; Sahajpal, R. Hansen, M.; Hurtt, G.: A global view of shifting cultivation: Recent, current, and future extent, *Plos One* 12 (2017), e0184479.

Highfield, R.; Coveney, P.: *The Arrow of Time*, Ebury Publ., 2015.

Höpp, G.: *Evolution der Sprache und Vernunft*, Springer, 1970.

Hossain, M.B.; Sultana, J.; Pingki, F.H.: Accumulation and contamination assessment of heavy metals in sediments of commercial aquaculture farms from a coastal area along the northern Bay of Bengal, Front. *Environ. Sci.* 11 (2023), 1148360.

Hoyle, F.: *A New Model of the Expanding Universe. Monthly Notices, Royal Astronomical Society.* 108 (1948), 372–382.

Hughes, A.D.: Defining nature-based solutions within the blue economy: The example of aquaculture, *Frontiers in Marine Science* 8 (2021), 711443.

Inwood, S.; Dale, V.H.: State of apps targeting management for sustainability of agricultural landscapes: A review, *Agronomy for Sustainable Development* 39 (2019).

Jackson, T.: The post-growth challenge: Secular stagnation, inequality and the limits to growth, *Ecol. Economics* 156 (2019), 236–246.

Johnston, J.F.; Allen, C.D.; Franklin, J.F. et al.: Changing disturbance regimes, ecological memory, and forest resilience, *Front. Ecol. Environ.* 14 (2016), 369–378.

Kaprelyants, A.S.; Gottschal, A.S.; Kell, D.B.: Dormancy in non-sporulating bacteria, *FEMS Microbiol. Lett.* 104 (1993), 271–286.

Kassi, N.J.K.; Decocq, G.: Successional patterns of plant species and community diversity in a semi-deciduous tropical forest under shifting cultivation, *J. Veget. Sci.* 19 (2008), 809–812.

Klijn, F.; UdoDeHaes, H.A.: A hierarchical approach to ecosystems and its implications for ecological land classification, *Landscape Ecology* 9 (1994), 89–104.

Koehler, J.M.: Symmetries in the genome structure of Bacillus subtilis shown by doublet frequency maps. *Proc. SPIE 4623* (2002), 59–66.

Köhler, J. M.: *Visualizing the Echo of Evolution: Tandem Repeats and Sequence Structure of Chromosomes of Saccharomyces cerevisiae shown by Doublet Frequency Distance Maps.* Preprints (2021), 2021080170.

Krug, H.-J.; Niedersen, U.: *Wilhelm Ostwald, Jahrbuch Selbstorganisation*, Halle/S., 1990.

Lambert, F.; Tagliabue, A.; Shaffer, G. et al.: Dust fluxes and iron fertilization in Holocene and last glacial maximum climates, *Geophys. Res. Lett.* 42 (2015), 6014–6023.

Loeffler, M.; Hinrichs, J.; Moß, K. et al.: Processing of biobased resources, in: *Bioeconomy* (ed. I. Lewandowski, Springer Open), 2018, 179.

Lovelock, J.; Sagan, L.: Atmospheric homeostasis by and for the biosphere: The Gaia hypothesis, *Tellus A* 26 (1974), 2–10.

Lyon, Ch.; Saupe, E.E.; Smith, Ch. J.; Hill, D.J. et al.: Climate change research and action must look beyond 2100, *Global Change Biology* 28 (2022), 349–361.

Maturana, H.R.; Varela, F.J.: *Autopoiesis and Cognition: The Realization of the Living*, Boston, 1980.

Margulis, L.: *Der Symbiotische Planet*, Frankfurt, 2018.

Meadows, D.H.; Meadows, D.L.; Randers, J.; Behrens, W.W. III: *The Limits of Growth: A Report for the Club of Rome's Project on the Predicament of Mankind*, New York, 1972.

Meadows, D.; Randers, J.; Meadows, D.: *Limits to Growth: The 30-Year Update*, London, 2006.

Meskhidze, N.; Chameides, W.L.; Nenes, A.: Dust and pollution: A recipe for enhanced ocean fertilization? *J. Geophys. Res. – Atmospheres* 110 (2005), D03301.

Mishima, K.; Nishimura, H.: Requirement analysis to promote smallsized E-waste collection from consumers, *Waste Management and Research* 34 (2016), 122–128.

Nakamasu, A.; Takahashi, G.; Kanbe, A.; Kondo, S.: Interactions between zebrafish pigment cells responsible for the generation of Turing patterns, *Proc. Natl. Acad. Sci. USA* 106 (2009), 8429–8434.

Nicolis, G.; Prigogine, I.: *Self Organization and Nonequilibrium Systems*, Wiley, 1977.

Pellizoni, L.; Centemeri, L.: Tackling material dependency in sustainability transition: Rationales and insights from the agriculture sector, *J. Environ. Policy and Planning* 24 (2022), 355–366.

Petersen, G.D.: Contagious disturbance, ecological memory, and the emergence of landscape pattern, *Ecosystems* 5 (2002), 329–338.

Pillay, T.V.R.: *Aquaculture: Principles and Practice*, Fishing News Bks., 1993.

Pyka, A.; Prettner, K.: in: *Bioeconomy* (ed. I. Lewandowski, Springer Open), 2018, 331.

Prigogine, I.: *Introduction to Thermodynamics of Irreversible Processes,* Charles C. Thomas Publishers, 1955.

Prigogine, I. and Glansdorff, P. *Thermodynamic Theory of Structure, Stability and Fluctuations*, John Wiley & Sons, 1971.

Prigogine, I.: *Vom Sein zum Werden*, Piper München/Zürich, 1992.

Razeghifard, R.: *Natural and Artificial Photosynthesis: Solar Power as an Energy Source*, Wiley, 2013.

Riesenfeld, E.H.: *Svante Arrhenius*, Leipzig, 1931.

Ronge, G.: Rudolf Clausius, ein Physiker des 19. *Jahrhunderts, Urania* 19 (1956), 231–238.

Rosenzweig, M.L.: Reconciliation ecology and the future of species diversity, *ORYX* 37 (2003), 194–205.

Ruiz, J.C.; Yanez, C.M.: The agrarian heritage: Definition, characterization and degree of representation in the UNESCO, *Boletin Asoc. Geograf. Espan.* 66 (2014), 105–124.

Santiago-Freijanes, J.J.; Mosquera-Losada, M-R.; Rosi-Diáz, M.; Ferreiro-Domínguez, N.; Pantera, A.; Aldrey, J.A.; Rigueiro-Rodriguez, A.: Global and European policies to foster agricultural sustainability: Agroforestry, *Agroforestry Systems* 95 (2021), 775–790.

Santoro, A.L.; Bastviken, D.; Gudasz, C.; Tranvik, L.; Enrich-Prast, A.: Dark carbon fixation: An important process in lake sediments, *PLOS ONE* 8 (2013), e65813.

Schimel, D.S.; Asner, G.P., Moorcroft, P.: Observing changing ecological diversity in the Anthropocene, *Front. Ecol. Environ.* 11 (2013), 129–137.

Schmithüsen, J.: Fliesengefüge der Landschaft" und "Ökotop", in: *Vorschläge zur begrifflichen Ordnung und zur Nomenklatur in der Landschaftsforschung. Berichte zur Deutschen Landeskunde 5*, Bad Godesberg, 1948, 74–83.

Sharp, P.M.; Averof, M.; Llooyd, A.T., Matassi, G.; Peden, J.F.: DNA-sequence evolution – the sound of silence. *Philos. Transact. Royal Soc. B – Biol. Sci.* 349 (1995), 241–247.

Smith, M.S.; Mbow, C.: Sustainability challenges: Agroforestry from the past into the future, *Curr. Opin. Environ. Sustain.* 6 (2014), V–VII.

Sprunger, C.D.; Martin, T.; Mann, M.: Systems with greater perenniality and crop diversity enhance soil biological health, *Agricult. Environ. Lett.* 5 (2020), e20030.

Stan-Lotter, H.: Halococcus dombrowskii sp. nov., an archaeal isolate from a Permian alpine salt deposit, *International Journal of Systematic and Evolutionary Microbiology* 52 (2002), 1807–1814.

Stiller, W.: *Ludwig Boltzmann*, Frankfurt, 1989.

Tanvir, U.T.; Zhang, J.; Canter, T.; Chen, D.; Lu, J.; Hu, Z.: Harnessing solar energy using phototrophic microorganisms: A sustainable pathway to bioenergy, biomaterials, and environmental solutions, *Renewable and Sustainable Energ. Rev.* 146 (2021), 111181.

Tommasi, F.; Mancini, L.: The issue of marine litter from a perspective of sustainable development for Mediterranean region: A review, *Fresen. Environ. Bull.* 28 (2019), 4963–4969.

Turing, A.: The chemical basis of morphogenesis, *Phil. Trans. R. Soc. (London)* B237 (1952), 37–72.

Varela, F.J.; Maturana, H.R.; Uribe, R.: Autopoiesis: The organization of living systems, its characterization and a model. *Biosystems* 5 (1974), 187–196.

Verma, J.; Petru, M.; Goel, S.: Cellulose based materials to accelerate the transition towards sustainability, *Ind. Crops and Products* 210 (2024), 118078.

Verreth, J.A.; Roy, K.; Turchini, G.M.: Circular bio-economy in aquaculture, *Reviews in Aquaculture* 15 (2023), 944–946.

Webster, K.: *The Circular Economy: A Wealth of Flows*, Allen Mac Arthur 2nd ed., 2017.

Whittaker, R. H.; Levin, S.A.; Root, R.B.: Niche, habitat, and ecotope, *American Naturalist* 107 (1973), 321–338.

Young, A.: Agroforestry, environment and sustainability, *Outlook on Agriculture* 19 (1990), 155–160.

Yu, Y.; Jin, K.; Chang, H.C.; Ling, F.F.; Barners, D.: A decision-making model for materials management of end-of-life electronic products, *J. Manufact. Sys.* 19 (2000), 94–107.

Zhabotinsky, A.M.: Periodic oxidation process of malonic acid containing solutions, *Biofisika* 9 (1964), 306–311.

Zhang, J.: Evolution by gene duplication: An update, *Trends in Ecology and Evolution* 18 (2003), 292–298.

Index

For Product Safety Concerns and Information please contact our EU representative GPSR@taylorandfrancis.com Taylor & Francis Verlag GmbH, Kaufingerstraße 24, 80331 München, Germany